9/25/92

ROTATIONAL PHYSICS:
THE PRINCIPLES OF ENERGY

ROTATIONAL PHYSICS

THE PRINCIPLES OF ENERGY

Myrna M. Milani/Brian R. Smith

FAINSHAW PRESS
WESTMORELAND, NEW HAMPSHIRE

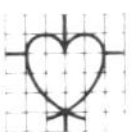

FIRST PRINTING

Text copyright © 1985 by Fainshaw Press

This book is manufactured in the United States of America. It is designed by James Brisson of Williamsville, Vermont, and published by Fainshaw Press, Olde Spofford Road, Westmoreland, New Hampshire 03467

Library of Congress Cataloging-in-Publication Data

Milani, Myrna M.
Rotational physics.

(Rotational physics and philosophy series)
1. Force and energy.   2. Physics.   3. Unified field theories.   4. Science—Philosophy. I. Smith, Brian R., 1939-   II. Title.   III. Series.
QC73.M47 1985   531'.6   85-16305
ISBN 0-943290-04-X
ISBN 0-943290-03-1 (pbk.)

0 9 8 7 6 5 4 3 2 1

Contents

Introduction

In this second book in the rotational physics and philosophy series, it's our purpose to lay the theoretical foundation for an entirely different approach to the collection of energy. Why a different approach? Because we *need* an entirely different approach. We all know Mother Nature isn't producing new resources such as coal or oil, but we want to eliminate the belief that the earth can use "bridge fuels" such as natural gas until something like nuclear fusion comes along. Both beliefs are as wasteful and inefficient as the processes they utilize. The time for the change is now, before we reach a point of no return, before the damage to the planet is irreversible.

Imagine a psychopath comes into your life and once a week cuts a piece of skin an inch square from somewhere on your body. At that rate, you can accomplish the necessary healing to keep you reasonably healthy. However, suppose our crazy visits you twice a week, then once a day, then every hour. Eventually a point is reached where you must flee, turn on your assailant, or die. So it is with our planet, a conscious organism with its own body, mind, and spirit.

In this book, as with *Primer*, we had a great deal of outside help. Albert Einstein, to whom the world is already indebted, gave us the most assistance with Gottfried Wilhelm von Leibnitz supplying some of the mathematics and Frank Lloyd Wright the architectural insights.

1 | A Review

In *A Primer of Rotational Physics* (Fainshaw Press, 1984) we presented some elementary material concerning the $_7\alpha$, the layers above the earth, and the various combinations of $_7\alpha$. Although *Primer* focused on a number of forms, two of primary interest were linear potential and the photon, both resulting from perpendicular bisecting $_7\alpha$ interactions. Let's review the first of these, the linear combination

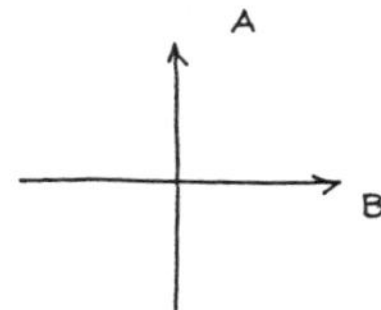

which we remember is actually

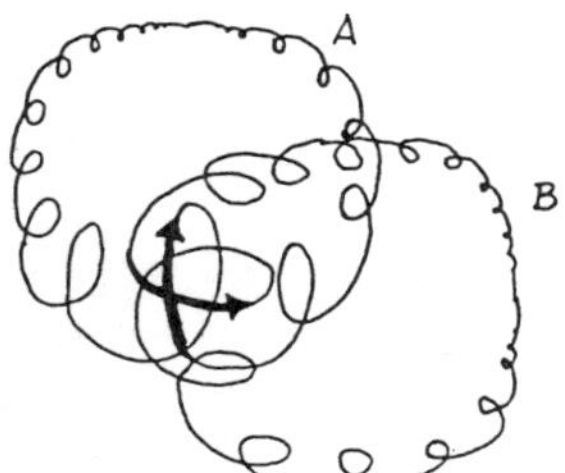

and represented mathematically as

$$\frac{V_{S_A}}{V_{L_A}} \Big/ \frac{V_{S_B}}{V_{L_B}} = \frac{0}{100} \Big/ \frac{0}{100}$$

We call this combined form pure potential. The single nomenclature 0/100 means that $_7\alpha$ is all linear velocity ($V_L = 100$ or 100%) and no spin (V_S = 0). The dual all-linear form predominates in outer space (layer 1) where the $_7\alpha$ have room to "stretch out" and the concentration of $_7\alpha$ is so low very little exists to hinder their linear movement.

The second combination—the photon and unit of light

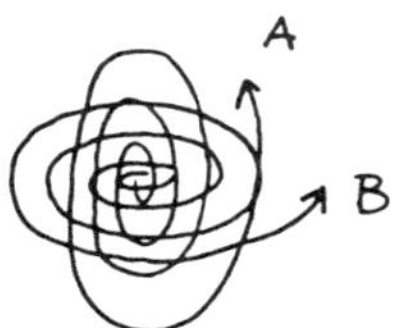

is represented as

$$\frac{V_{SA}}{V_{LA}} \bigg/ \frac{V_{SB}}{V_{LB}} = \frac{50}{50} \bigg/ \frac{50}{50}$$

Photons come in different "sizes", many of which are beyond our perceptual limits, being either too large or too small to see:

sort of like Goldilocks and the three bears.

Even if we have one $_7\alpha$ photon of the proper perceptual size we need 10^{13} photons per cubic millimeter (mm^3) for most of us (80%) to see it.

Because any $_7\alpha$ can assume any configuration in the continuum

we can have an infinite variety of forms (3/97, 78/22, 41/59) and therefore an infinite variety of intersections. So if

$$\frac{0}{100}\Big/\frac{0}{100}$$

represents two all-linear moieties and

$$\frac{50}{50}\Big/\frac{50}{50}$$

two half-linear velocity, half-spin entities, we can have many intermediate forms of like $_7\alpha$:

$$\frac{10}{90}\Big/\frac{10}{90}, \quad \frac{25}{75}\Big/\frac{25}{75}, \quad \frac{40}{60}\Big/\frac{40}{60}$$

as well as unlike (unequal) $_7\alpha$ such as

$$\frac{10}{90}\Big/\frac{25}{75}, \quad \frac{40}{60}\Big/\frac{10}{90}, \quad \frac{21}{79}\Big/\frac{7}{93}$$

These forms aren't "pure" but represent combinations which function somewhere between potential and light. Imagine what happens when you take each of two wires from the poles ($+$ and $-$) of a car battery and put them together—BAM! You get electrical flow (potential) as well as sparks (light).

All of the forms of "energy" we *presently* use are impure or mixed forms. As voltage flows within a wire it also produces a small amount of heat (waste) due to frictional losses within the wire. A light bulb whose function is to produce illumination also throws off heat: more waste. A photovoltaic cell converts some light to electricity but not efficiently. One objective of rotational physics is to establish the theory behind the *pure* forms of energy—potential, light, heat, sound, and pressure—so devices may be constructed which use these forms. These machines and processes

- Use no natural (or unnatural) resources.
- Are safe.
- Produce no waste.

Humankind's greatest single error in its entire history on the planet is extrapolation—more of the same. Rather than attempt to explain energy and mass in a way that transcends primitive experience, the human race in its quest for more and more energy, simply copies the caveman's fire. In two hundred years about all we've done is move fire from the open (fireplace) to a metal container (the Franklin stove, the oil burner) to a total abomination (the nuclear fission reactor). Now some are even attempting to create a "fire" whose temperature will be in millions of degrees centigrade—the fusion reactor. All of this activity invariably leads to storage, storage leads to greed, and the result of storage and greed is conflict and war.

The evolution of our weapons parallels our energy development. Primitive man fought with sticks and stones; then followed the spear, the arrow, and the gun, each one separating the combatants by a greater distance. World leaders in their final act of cowardice can now command nuclear-tipped missiles to rain death on millions on the other side of the globe. A "hit" by an offensive nation is recorded as a spot on a computer screen: No human witnesses the unleashed carnage. Both the U.S. and the U.S.S.R. have contemplated "doomsday machines"—computers which will launch missiles even if one country or the other is already totally devastated. It would be interesting to know whether computers with such technological know-how can answer the simple question, "Why?"

As we said in *Primer,* we can't expect our leaders, our governments, or our large institutions to solve these problems. It must come from each and every one of us. The time has come to make a personal stand, and it may not be easy. Those who work in the so-called defense industries are intuitively aware that their efforts don't benefit the human or any species. Those who pull oil from the ground, those who denude our forests on a wholesale basis, those who fill our atmosphere with the sulfurous fumes creating acid rain are all similarly aware of "What does it profit a man. . . ?" So the time has come to say, *Stop.*

To continue our journey together so we may learn and do, let's first examine the basic forms of energy. Although the material in this book can more or less stand on its own, you may want to review the preliminary material presented in *Primer.*

2 | The Energy Forms

Because the majority of this book deals with the principles of energy, it's important to establish the nature of the various energy forms at the $_7\alpha$ level. As you read this chapter, bear in mind that when we refer to a form as heat or sound, these aren't the same forms of heat and sound we know from our experience. For instance, we spoke of pure linear potential

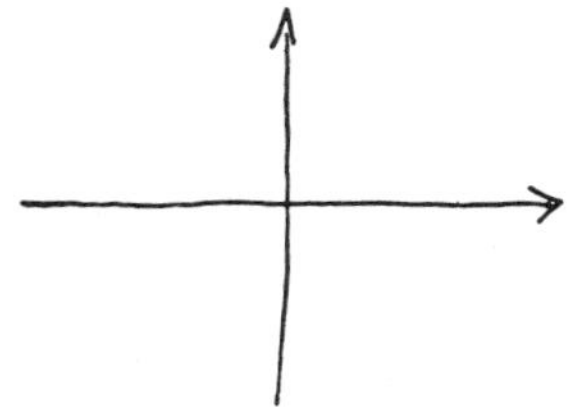

resulting from the perpendicular, bilateral intersection of two equal, all-linear-velocity $_7\alpha$, yet we also know that in electricity, voltage is often called potential. Is our linear potential the same as electrical potential? No. Although there are many differences, the most meaningful one is how the two potentials differ in their effects on us. If you grab two wires from a normal house wiring system, the result is shocking indeed. It would certainly be painful and might even be fatal. On the other hand, if your local utility could supply pure potential and you connect yourself into the circuit, the result would be exhilarating, not uncomfortable. In many ways the differences between pure potential and voltage or potential resulting from current energy-producing technology are like those between pure and muddy water. To be sure, we can design systems with special pipes and filtering mechanisms to allow for or eliminate impurities, but it would be far more efficient to begin with pure water. Similarly we could drink muddy water, but the impurities could be just as detrimental to our health as the impure voltage.

Why don't we use a different word or words to describe potential? Like muddy and pure water share certain properties, so pure energy forms are similar to those that surround us every day. Rather than refer to pure potential, light, heat, sound, pressure, as "rigoflex", "aardvark", "demento", "jurgistan" and "flymap", we feel it's less confusing to use the same terminology. Think of the potential, light, heat, sound and pressure you're used to as a double or stand-in for a movie star. The stand-in may look like John Wayne, may perform his lines and actions, and may even be referred to by the director and producer by the same character's name the Duke plays in the movie—but the two men are not the same. Like muddy water and the energy forms we're used to, the stand-in can do a lot of things John Wayne can do, even look like him, but he can never *be* John Wayne.

Polarity

In order to understand pure energy we need to understand the concept of polarity. In *Primer* we said that if we have a linear system in motion,

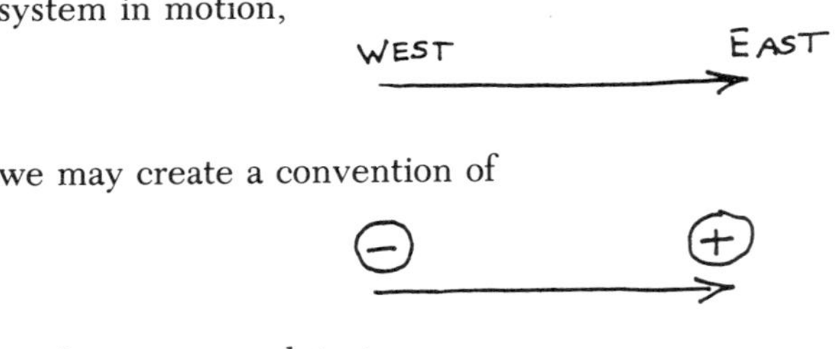

we may create a convention of

or, in more complete terms:

The mid-section $(\pm)$ is merely the area of neutrality in which negativity gives way to positivity.

However, because the $_7\alpha$ is rotational rather than linear, it creates a circular (actually spiral) system having two areas of neutrality:

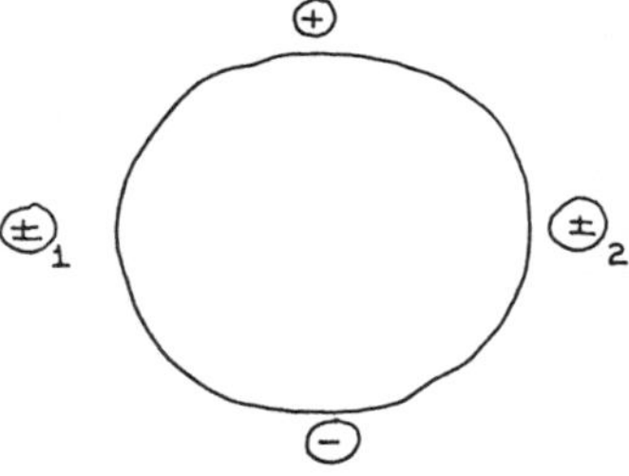

When two $_7\alpha$ intersect, the possible combinations are indeed infinite, but let's consider the major ones. Because there are four states of our $_7\alpha$ ($\pm_1$, $\pm_2$, $+$, $-$), we can develop 16 combinations of intersections. Note that $\pm_1$ is different from $\pm_2$ because the former represents the intermediate state going from minus ($-$) to positive ($+$) and the latter the state from ($+$) to ($-$).

Now we may construct our table of probable intersections of two $_7\alpha$, A and B:

$_7\alpha$ B \ $_7\alpha$ A	$+$	$\pm$	$-$	$\pm$
$+$	$\dfrac{+}{+}$	$\dfrac{+}{\pm}$	$\dfrac{+}{-}$	$\dfrac{+}{\pm}$
$\pm$	$\dfrac{\pm}{+}$	$\dfrac{\pm}{\pm}$	$\dfrac{\pm}{-}$	$\dfrac{\pm}{\pm}$
$-$	$\dfrac{-}{+}$	$\dfrac{-}{\pm}$	$\dfrac{-}{-}$	$\dfrac{-}{\pm}$
$\pm$	$\dfrac{\pm}{+}$	$\dfrac{\pm}{\pm}$	$\dfrac{\pm}{-}$	$\dfrac{\pm}{\pm}$

Let's define how we're going to use the symbols $+$, $-$, and $\pm$. Linear velocity (V_L) by definition has direction: It's the word "velocity" that does it because velocity implies both magnitude and direction. This is what's known in science as a *scalar* quantity, compared to speed which is *planar* and has magnitude only. For example, if you're driving your car at 70 km/hr, that's your speed; if you're traveling at 70 km/hr due east, that is your velocity.

Because linear velocity has direction, it's moving someplace compared to pure spin which occurs in place. Therefore, we shall define V_L as positive $(+)$ and V_S as negative $(-)$. The $\pm$ state represents an equal probability of the $_7\alpha$ functioning as linear velocity and spin. In other words,

a. The $+$ state is all linear velocity, thus

$$\frac{V_S}{V_L} = \frac{0}{100}$$

b. The $-$ state is all spin and

$$\frac{V_S}{V_L} = \frac{100}{0}$$

c. The $\pm$ state is

$$\frac{V_S}{V_L} = \frac{50}{50}$$

We may now reconstruct our previous table in the following way:

$_7\alpha$ B ╲ $_7\alpha$ A	$\dfrac{0}{100}$	$\dfrac{50}{50}$	$\dfrac{100}{0}$	$\dfrac{50}{50}$
$\dfrac{0}{100}$	$\dfrac{0}{100}\Big/\dfrac{0}{100}$	$\dfrac{0}{100}\Big/\dfrac{50}{50}$	$\dfrac{0}{100}\Big/\dfrac{100}{0}$	$\dfrac{0}{100}\Big/\dfrac{50}{50}$
$\dfrac{50}{50}$	$\dfrac{50}{50}\Big/\dfrac{0}{100}$	$\dfrac{50}{50}\Big/\dfrac{50}{50}$	$\dfrac{50}{50}\Big/\dfrac{100}{0}$	$\dfrac{50}{50}\Big/\dfrac{50}{50}$
$\dfrac{100}{0}$	$\dfrac{100}{0}\Big/\dfrac{0}{100}$	$\dfrac{100}{0}\Big/\dfrac{50}{50}$	$\dfrac{100}{0}\Big/\dfrac{100}{0}$	$\dfrac{100}{0}\Big/\dfrac{50}{50}$
$\dfrac{50}{50}$	$\dfrac{50}{50}\Big/\dfrac{0}{100}$	$\dfrac{50}{50}\Big/\dfrac{50}{50}$	$\dfrac{50}{50}\Big/\dfrac{100}{0}$	$\dfrac{50}{50}\Big/\dfrac{50}{50}$

Combined Forms

For those who followed the developments in *Primer*, you may remember that the photon, the unit of light, is represented mathematically as

$$\frac{V_S}{V_L} = \frac{50}{50}$$

for each of the $_7\alpha$ components and

$$\frac{50}{50}\bigg/\frac{50}{50}$$

for the inherent pair. Graphically, our photon looks like this:

Let's pick another pair of $_7\alpha$ from the table and examine its form, both mathematically and graphically:

$$\frac{100}{0}\bigg/\frac{50}{50}$$

Here one $_7\alpha$ is all spin ($V_S = 100$) and no linear velocity ($V_L = 0$) and the second is a 50/50. We can draw this as

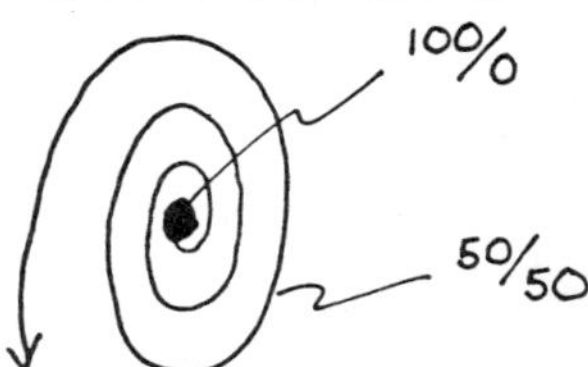

Note that this is also the same as

$$\frac{50}{50}\bigg/\frac{100}{0}$$

or

This form is pressure.

If we tabulate our probabilities we have

$Qty.$		$V_{S_A}/V_{L_A} \big/ V_{S_B}/V_{L_B}$	$Symbol$
1	—	$\%_{100} \big/ \%_{100}$	
1	—	$100\% \big/ 100\%$	
2	—	$\%_{100} \big/ 50\!/\!50$	
2	—	$50\!/\!50 \big/ \%_{100}$	
2	—	$100\% \big/ 50\!/\!50$	
2	—	$50\!/\!50 \big/ 100\%$	
4	—	$50\!/\!50 \big/ 50\!/\!50$	
1	—	$\%_{100} \big/ 100\%$	
1	—	$100\% \big/ \%_{100}$	

Condensing the above nine configurations, we discover there are six forms of pure energy:

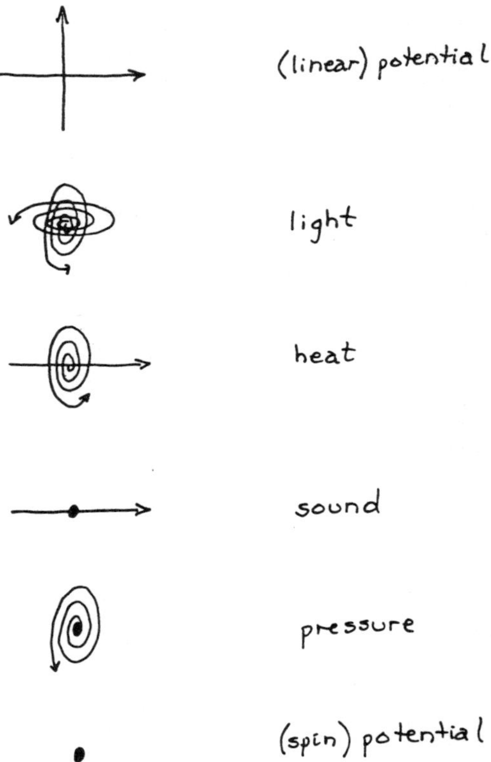

And all this seems to make some sense. Pressure and sound are energy functions of mass-ness. Sound needs mass (spin) in order to exist, i.e., to travel. We create increasing pressure by compacting mass. Heat is nothing more than fast sound-motion plus less spin. Direct high energy sound at an object—ultrasonic cleaning, for example—and you have heat, especially when the object cannot move. Light, on the other hand, is slow potential.

We're not going to develop the theory behind the energy forms any further in this book for one very strong reason: For a long time technology has far outstripped the philosophy that should accompany it. All we need do is to look at the creation and subsequent use and deployment of nuclear weapons for evidence of this phenomenon. Therefore, rather than present designs for devices that can create the energy forms, we're going to devote this book to the development of the underlying philosophy. How new is this philosophy? Among other sources, Christ's teachings contained the basic working philosophy of rotational physics, but a good deal of it was lost or ignored for various reasons. The philosophy, *per se,* isn't new; how we intend to use it is.

Current Energy Problems and Beliefs

Let's look at some of the present problems and beliefs that exist with energy creation, distribution, and use. Such awareness provides a solid base from which to begin because any new approach must seek to alleviate all or at least a majority of our current energy-related problems.

- Our energy sources aren't endless. If we continue the way we're going, someday we're going to extract the last barrel of oil from the ground. Even high-grade uranium ore is getting more difficult to find.
- All current energy systems produce waste in one form or another. Most of it is detrimental and hazardous.
- Large energy-producing facilities are expensive to construct and run. A nuclear plant may cost billions and take years to build.
- Distribution networks, lines and poles, require more resources.
- Current systems are highly inefficient; we never get out what we put in.
- There are constant factors at work that can negatively affect energy distribution—availability of fuels, weather, strikes, brownouts, insolvency of utilities.
- It isn't safe. Almost all generating stations produce electricity

which is lethal to humans. Three Mile Island graphically demonstrated the danger associated with nuclear plants and unfortunately we'll have more of these catastrophies until we fully realize what we created.

- All systems, from generators to wires, have a limited life and must eventually be replaced.
- People worry about it—"What happens if I lose power in the winter?"
- The average consumer has little to say about what goes on in the utilities. Very few people in the Western world have chosen to become self-sufficient in terms of energy usage.
- It can probably be said that most humans believe they can't live without "man-made" energy. Even the most primitive peoples use fire for cooking.
- In large producing plants, especially the nuclear variety, the cost of safety measures often exceeds the cost of the energy-producing mechanisms. Add to that the effort and cost of satisfying one regulatory agency after another and it all hardly seems worthwhile.

The way to overcome these difficulties is not to rage against them, but to work towards new solutions; and the following philosophy will help lay the mental and spiritual groundwork for these solutions. We can no longer build our houses on sand or our energy-producing facilities on groundless foundations.

Creating Change

We must take great care during the coming decades of change, more care than we've ever exercised as a population in the past. We must proceed from what we know to what we don't know, from the simple to the complex because when we say that it's possible to create devices that operate counter to existing theory and practice, the potential exists for great fear and the resulting antagonism. When we introduce designs for new mechanisms and processes that produce one of the energy forms, heat for example, remember that the form being discussed is different from that normally thought of.

We're postulating a new form of energy that functions *like* heat, or sound, or pressure which thereby frees this theory from the limitations imposed by some very narrow definitions in the past.

Suppose someone brings a new species of animal to you that has some but not all of the properties of a cat. If you are a cat lover, own several cats whom you enjoy and are quite attached to, and your friend says, "Here is a kind of cat that is much better than your cats—Get rid of your cats right now," imagine how defensive you would be. However, if your friend says, "Here's a new species of animal that might interest you. Although it's mostly cat-like, it can display characteristics of a dog, or even a bird. Keep it awhile and see what you think." Now you have been presented with something similar yet also quite different from what you know. Your beliefs haven't been challenged; what you currently have isn't being denigrated, and therefore neither are you. Chances are given your own opportunity to explore and study this new creature, you will decide it is a far superior companion, particularly because it *immediately* fits it well with your present cat population. Accepting the new animal into your home creates no disharmony in you, your cats, or your lifestyle and does not force you to get rid of your other pets.

This is a crucial point for regardless how superior the new species may be, if we *must* give up the familiar to accept it, most will resist. However in our example it isn't necessary to abandon the familiar to accept the new. This new animal is quite loving and compatible with the others and at first supplements their activities while simultaneously eating much less, requiring less personal attention, and producing much less waste. Eventually you discover that as your other cats age and die you feel no need to replace them; the new animal is quite capable of filling your needs.

Think for just a moment of our long-standing relationship with energy. Even though we realize that the system lacks something, right now it's all we know. It's familiar and predictable even in its worst aspects. However, even though few like being married to a system of wires, receptacles, switches, transformers and cables that he or she doubts, having an outsider sever that relationship is intolerable. And when do we *most* resent outside intervention? Surprisingly, when the change "forced" on us is *most* aligned with our own beliefs. Because someone suggests the obvious, we feel inferior because we didn't respond to our own needs first—and we lash out in anger.

This is why this book is as much or even more philosophical than scientific, although the two can't really be separated any longer. In relation to the coming changes, realize that no one (NO ONE!) can ever make you do, think, or believe anything. The changes will come, but slowly enough that we can all deal with them, each of us in our own time. No structure will be toppled, no government radically altered, no organization changed whose time for change has not naturally come.

Now let's devote our energies to the principles, beginning with an overview.

3 | The Principles Of Energy: An Overview

Principles aren't like laws—they're less absolute, more like guidelines. In this chapter we're going to propose twelve scientific and philosophical principles to guide us through the following discussion of rotational physics.

The First Principle

The First Principle: If energy is created at the smallest level $(_7\alpha)$, there's *no* waste. All current energy-producing systems create waste. The burning of wood and coal creates useless ash and light; the oxidation of liquids (oil, gasoline) and gasses (natural gas, propane) produce waste such as carbon monoxide, unburned hydrocarbons, and light. Lamps produce unwanted heat. Nuclear reactors create numerous wastes so dangerous they must be buried and heavily shielded to protect all living things from their negative effects, and even then we're not really sure.

When what we see as mass is totally converted to energy, there's no waste. However, there is a telltale marker to let us know when such a conversion is complete. What is this marker? To answer that question, let's take another look at a familiar equation,

$$e = mc^2$$

If a mass-to-energy conversion is truly complete and all mass becomes energy, the mass must equal the energy in such a process, or

$$e = m$$

If e = m, then we can substitute e for m in the equation $e = mc^2$:

$$e = ec^2$$

$$\frac{e}{e} = c^2$$

reversing

$$c^2 = \frac{e}{e} = 1$$

and

$$c = \pm 1$$

Similarly, we may substitute m for e in the equation ($m = mc^2$) and get the same results, $c = \pm 1$. Plus or minus one what? One *photon*. The $\pm$ denotes the photon as the dual 50/50 $_7\alpha$ configuration. If a mass is totally converted to energy, the only thing we see is one photon.

To be sure, traditional non-rotational laws of algebra and physics say we can't do what we just did. However, when we use rotational mathematics such relationships are quite valid. For now, simply remember that one proof of a clean or total mass-to-energy conversion is the existence of a single photon at the end of that reaction.

So, it's possible to create a clean and total mass-to-energy conversion. Because clean non-waste producing mass-to-energy conversions don't exist in current technology, how do we know they're not only possible but highly probable? As always, our best proof comes from nature and ourselves. We may view the death of any living being as a mass-to-energy conversion. Unhampered by embalming fluid and cement vaults, a dead animal or plant converts (reverts) to energy quite nicely. Or think of yourself in a state of extreme anger or tension (i.e., energy-ness); once you do something physical—run around the block, clean the attic, jump up and down and sob hysterically—that is *equal* to that feeling, it disappears.

The Second Principle

The Second Principle: Process is less important than result. Said another way, "Create what is needed." If we want light, we should create light, concentrating only on the final conversion and outcome, not the process. Don't think, "To make light, I take water, heat it with oil until it's high pressure steam, which then turns a steam turbine whose output drives a generator to supply electrical power which travels along wires, then excites a filament in a bulb producing light when I flip the switch."

Let's look at all the different forms of energy created during this multi-step process and what becomes of them. The burning oil creates heat and light but only the heat is used to create pressure; the light is a waste product. The pressure is used to turn the turbine but when the turbine turns, it also produces unneeded heat (friction) as well as the spin necessary to power the generator. In addition to creating the desired electrical potential, the generator also produces more losses which are added to our growing stock of waste. This potential is forced through a wire as voltage producing waste heat, finally creating light—and yet another ration of unneeded heat. In the process of creating light we created heat, pressure, sound and potential discarding all but a fraction of the potential.

Compare this to merely thinking, "I want light to fill this room," and then creating the necessary changes at the 7α level to produce that light. Cause has little importance in the proper use of energy; it's the effect we should focus on. Defining the "how", the cause or process often introduces needless energy-wasting steps. It's as though upon proving electricity produces light, we feel obligated to build a step-wise system to convince the rest of the world this is the way things work. Pretty soon such systems come to be viewed as the *only* systems that work.

Think about youngsters learning to walk. If we look at a group of children achieving this skill, we see each child uses his or her own method. Some may sit quietly and work through all the mental and emotional aspects, then suddenly get up and walk. Others may combine thought, emotion and physical activity, little faces tense with concentration as they make one or two tentative steps. Others may think, make a mad dash between chair and table, then collapse in laughter or tears depending on their emotional evaluation of their performance.

Suppose fifty years ago a scientist devised a process for teaching children to walk which produced that ability in 98% of the youngsters taught in accord with this method. It would be very easy to say that was the best, perhaps even the only *correct* way to produce this skill in children. Pretty soon we could even come to believe children couldn't walk, or at least not walk *properly*, unless that method were used. In other words, the process could become more important than the result. Ancient sages understood this lesser position of process when they noted "the end justifies the means." Unfortunately in our unwillingness to accept this as true on the most personal and individual level, we often interpret this as a most negative rather than a basically unemotional statement of fact.

The Third Principle

The Third Principle: Heat-based reactions are wasteful and inefficient. Because $_7\alpha$ reactions proceed without the addition of heat, temperature dependency in energy creation is unnecessary. Suppose we want to create linear potential: we bring two linear $_7\alpha$ together

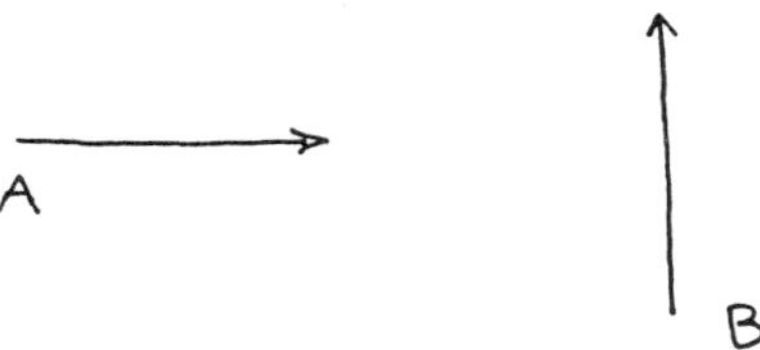

so they form the perpendicularly bisecting pair characteristic of pure potential:

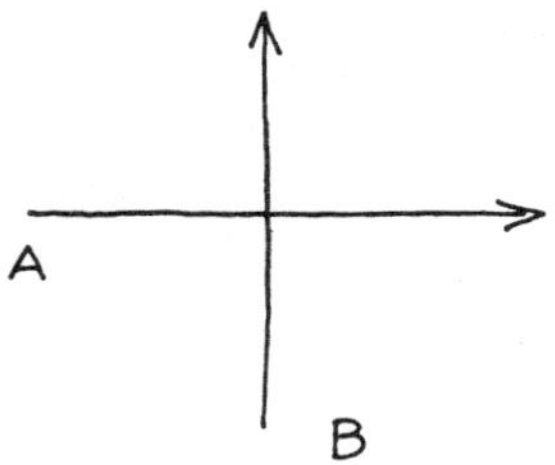

But what if something (heat, for example) is in the way?

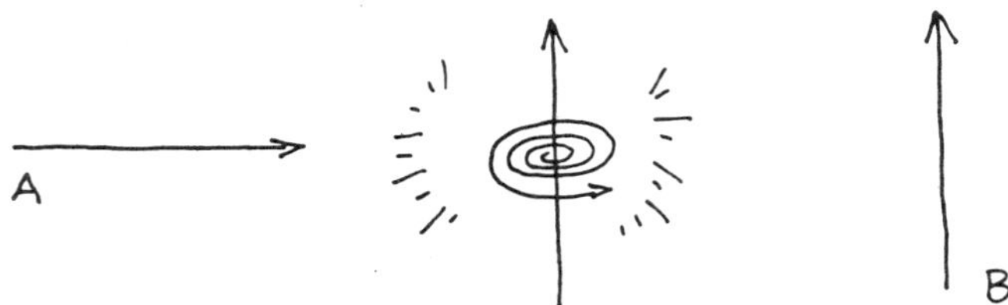

We either have to remove the heat or drive A into B *through* the heat configuration which requires more energy. One reason current processes and reactions are so inefficient is because they must always work to overcome the effects of heat.

A valid test for the creation/conversion efficiency of any energy system is how much energy it takes to create that energy, including any human energy expended. A major question when a nuclear reactor is decommissioned is "Did it produce more energy than it used?" Usually the answer is "No", or "Just barely". Much of this inefficiency occurs because nuclear fission is a high-heat reaction. The heat isn't produced directly but is actually a waste product and very little of it has the pure $_7\alpha$ heat configuration:

Much of it is in intermediate forms such as

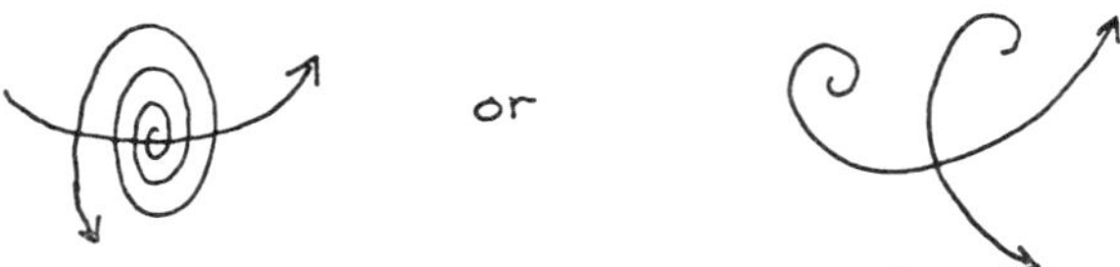

In human terms, think of two individuals who have different ideas about making this a better world. One may feel the best way is for some central power to provide the best for all while the other feels each individual must provide for him or herself. If we merge these two beliefs, we realize that *together* they speak to the needs of all people and as such could indeed make this a better world.

However if these beliefs are separated by the heat of anger (fear), the final result relative to making this a better world is minimal. Making a world safe for democracy is only a valid form of energy for those who *want* a democracy; for those who prefer a monarchy, such energy doesn't fulfill their needs. However, if both factors waste time and energy fighting each other, they usually discover their followers turn to other forms of energy and support.

The Fourth Principle

The Fourth Principle: Any chain reaction is unnatural. By chain reaction we refer primarily to nuclear reactions where uncharged particles (neutrons) slam into already unstable nuclei of (unnatural) uranium and plutonium at an unnatural rate causing them to separate into radioactive isotopes of lighter elements and emit more neutrons. Such reactions don't occur in nature.

Not only are the components in these reactions unnatural, but also their timing. Although the numbers of reacting elements may increase exponentially in nature, in general the time interval for those increases remains constant. If it takes a single cell twenty-four hours to divide into two new cells, it also takes each one of those cells twenty-four hours to divide. If the average woman's first fetus requires a nine month gestation period, her second, third or fourth child normally also requires this same interval. Compare this to unnatural chain reactions where the time interval decreases as the numbers involved increase. Think of yourself ascending and descending a steep incline on a roller coaster. First you creep along covering little distance per unit time; then the pace picks up and each second sees more and more feet of track slip by. Exciting perhaps, but how *natural* does this feel? Is it the kind of feeling you would like to feel day in and day out? Do you think your body's designed to recognize that state and its effects as normal?

What occurs in nature are simultaneous, contingent events. Anything, any reaction that can't be easily controlled and run to completion, or that creates waste, is unnatural and can't respond to the laws of nature. It is, in fact, out of synchrony with the universe.

The Fifth Principle

The Fifth Principle: Energy in one form can supply all the needs of the earth. This isn't a particularly surprising statement because electrical energy is presently used to provide light, heat resistive elements (such as toasters and baseboard heaters), turn a motor, power a TV set. The problem is that the creation of this form of electrical energy produces waste and consumes resources not only in its process but in the equipment needed for those processes. The steel, copper, and aluminum hydroelectric generator powered by rushing water must eventually be replaced, and this replacement requires energy.

It's possible to collect linear potential (Think of it as a very pure form of electricity.) to supply our needs. We can't deplete this supply of linear $_7\alpha$ forms because the number available is infinite. Because the layers above the earth and beneath its surface contain a fixed number of $_7\alpha$ and are self-replenishing, we can't exhaust any layer.

Does the idea of using one form of energy to supply all your needs bother you? Think of all the people who believe money, fame, or power will solve all their problems; in their rather limited way they're expressing a belief that a lot of solutions can evolve from one basic form. Better yet, look in the mirror: Inherent in each one of us is every possible form of solution. We can perform an infinite variety of physical, mental and emotional acts in response to an infinite number of needs—all that potential is there in that one being, that one and only form on this earth that is "pure" you.

The importance of this principle isn't that one form *must* supply all needs, but that it *can*. Although pure linear potential can supply all our needs, other pure energy forms such as light, heat, sound, pressure can also fulfill our needs.

The Sixth Principle

The Sixth Principle: The more specific the energy-creation device, the more energy it wastes. Specific energy-creating devices (oil burners, electrical generators, nuclear reactors) waste energy in two ways:

- Their conversion efficiencies are low, in some cases less than 10%.
- They require more energy to build and run than they produce.

For example, the atmosphere surrounding your home and yard abounds with all forms of energy. If you install a system which *only* collects light, heat, pressure (wind), or any other form, all other forms are not only not utilized, they must be blocked from the system. This requires the energy-consuming creation of more extensive collection devices.

Think of hiring an administrative assistant who can only do marketing research versus one who is adept in all aspects of business management. Not only can't your exclusive marketing researcher do other work for you, that person must be maintained (paid) even when there is nothing for him or her to do.

The ideal collector is a totally passive system. If we have a system that collects *all* forms of energy, then we can create the amount and form of energy we need, rather than creating our needs to fit the amount and form of energy available.

The Seventh Principle

The Seventh Principle: Whenever energy is stored, more energy is lost than gained. Believing that storage is necessary leads to distrust, weaponry, and wars. The problems associated with storage have been well known for years. One of the Ten Commandments advises us not to be envious of or covet what someone else has and Christians are admonished not to "store up treasures on earth". If each person, village or country creates *only* that energy which is needed *in situ* (*where* it's needed) at any given moment and no more or less, then there's no buildup, no stockpiling, no need for envy.

In situ energy production by all who need it versus production and storage by a few is, in many ways, comparable to each person having one apple tree containing all varieties ripening at all times versus the existence of one large orchard in the state. Although we may say it is more efficient to cultivate apples from hundreds or thousands of trees versus a single one, we get a completely different view if we examine the *needs* of the single individual. If we have a tree in the yard we can choose where to plant it, how to cultivate it (organically or with chemical fertilizers and insecticides, for example), when to pick the apples, and how many and what apples to pick. If there's only one orchard in the state, we must either travel there or set up some system whereby the apples are shipped directly or indirectly to us. If we go to a store, we often discover the kind we prefer only come in five pound bags, far too many for a single person to consume at one time. If we buy our apples singly, we discover they are quite expensive. Whether we go to the store or the orchard we not only consume energy in the process, we often discover we must tailor our needs to fit what's available. With our multi-variety single tree, our needs are much more likely to be met when we want and in the desired form. Even if we can't eat or otherwise use all the apples produced by our single tree, those that fall add their nutrients to the ground and nourish the tree.

This example doesn't mean we advocate total self-sufficiency for everyone. However when we depend on something as much as we depend on energy, it seems infinitely more logical and efficient for each individual to have the capacity to produce his or her own rather than being dependent on energy-consuming wires leading from sources miles away that employ incomprehensible processes and create great waste.

It's never the creation of sufficient energy to fulfill one's needs *and no more* that creates problems—it's the excesses, their creation and storage that creates waste. More people can be gainfully employed filling the needs of all than are now being employed to over-satiate a few, frustrate the majority, and intimidate a small, but increasingly significant and powerful minority. This isn't to say that *in situ* energy production is costly in time or labor, but that it opens unlimited resources in areas previously considered inaccessible or uneconomical. *In situ* energy production gives all we need and leaves nothing more than footprints.

Stored energy in any form can't maintain that state for long and is in a constant process of reverting to mass (that which we see, hear, smell, taste, or feel). Free energy is always looking for a way to manifest itself in this, our mass-oriented reality. As with a child, it's always wiser to provide valid channels for energy expression than let the source go unguided and unused.

The current processes for creating energy create so much waste the end-product, in both amount and form, is remote from the energy potential within the original mass.

Like the child forced to go through a process of walking which he or she believes unnecessary or unnatural may incorrectly impress the teacher as being lazy or of less than average ability, so the energy produced by current nuclear technology is barely representative of the potential inherent in the pure form. If we go from the point of energy conversion directly to the point of need, the energy created has greater power than even suspected in the past.

The Eighth Principle

The Eighth Principle: Mass is both created and destroyed. This principle may seem to contradict what we just said about energy but energy, although changeable, is more "permanent" than mass. Mass is impermanent. Your physical form is impermanent; it only lasts a certain number of years in accordance with individual and majority beliefs. Your spirit, your energy-essence if you will, is permanent. As long as your energy-essence balances your physical form, you exist on this earth. Once the energy form departs from the body, the body begins to decompose and "disappear".

Think of a clump of daisies growing beside the road. The flowers, stem, leaves and roots make up the mass form of the plant. However, we know that form depends on at least one form of energy, that from sunlight, for its existence. If we place the plant in total darkness, even with the finest soil, temperature and moisture control, it dies. So the plant is totally dependent on the sun's energy even though it doesn't resemble the sun in form at all. Furthermore, the sun's energy manifests itself in many other forms (some which are even antagonistic towards the daisy), and is present preceding and following the daisy's lifespan. The energy is always there becoming incorporated into one form of mass or another, existing before the creation and after its demise, as well as during its existence here.

We can't see or measure energy except in terms of form. As soon as we measure it, energy ceases to exist *at that point*. Similarly, when mass is converted to energy, it ceases to exist. If a house is burned down, hasn't it been converted to measurable light and heat energy? Within our usual, classical thinking that's true. However, if we think carefully about the actual "house" state prior to ignition and the *measurement* of its luminous and thermal energy following the fire, there is essentially nothing, or more correctly, no "houseness" left. What is left is debris and measurements. The house is form previous to the fire; the subsequent measurement of heat and light produces a *different* kind of form—changes in the meters, dials and digital readouts of instruments—which then permanently change the form of these instruments.

In high school physics we learned that "matter" cannot be created or destroyed, merely changed from one form to another, and "matter" was equated with mass. If your reaction was like ours, you may have sensed that intuitively the statement just didn't seem right. A baby is born, an old person dies. We know the fetus existed in the cells of sperm and egg and intuitively feel the potential human didn't just appear there as a result of parental physiology. Similarly we recognize death in the *form* of the individual in the casket at the funeral home, but we don't recognize that as the total of *all* we recognize as Aunt Harriet or Uncle Fred. Their *mass* form will indeed disappear just as that of an infant miraculously appears. The mass form is created and destroyed; the energy goes on forever. Mass can indeed be created as it must be when we consider the probable nature of the $_7\alpha$. In like manner, mass is destroyed as well, going quite literally out of existence: out of our reality.

The Ninth Principle

The Ninth Principle: For every (re)action occurring in nature, some change in either mass or energy occurs. Everything in the universe *wants* to be at one. It's a state constantly sought but achieved only temporarily—a transition point or interval. Because everything in the universe is changing and because everything is connected, there are constant "pulls" on the at-one state. Therefore, reactions occur at many levels and these go hand-in-hand with energy and mass changes. Think of the at-one state as the fulcrum of a teeter-totter with energy on one side and mass on the other. Any time there is a change on either side of the seesaw, it rocks or rotates on the fulcrum trying to re-establish its balance. Although it isn't part of the components inter-reacting at either end, the fulcrum is critical in the overall motion and balance of the system.

This isn't simply a restatement of Sir Isaac Newton's principle, "For every action there is an equal and opposite reaction". "Opposite" reactions aren't possible because there are no such things as opposites. The "up" end of a seesaw by itself has no meaning. An opposite is merely the reflection of an action in another plane of reality, not a form that is completely different but rather the other "half" of a "whole". The reason positive and negative ions, poles, or charges attract each other is because the parts want to be whole, to be one, just as all things want to be at one. In our discussion of the First Principle we spoke of the presence of a single photon as indicative of the cleanest reaction. The presence of this one photon signaling the complete conversion of one form to another is no different from what happens on our seesaw. If both ends are equal, the downward motion of one is balanced by the upward motion of the other. Although this opposite motion may continue for a while, because they are equal the potential exists for them also to be (at) one; if $a = b$ then $a/b = 1$ and the system soon stops. Although this position is most stable, anyone who's ever played on a seesaw knows it's also quite boring. Although it's nice to know the potential for the (at) one state always exists, it's also nice to know the inter-relationship of all that is is constantly changing the arc or rotational character of our infinite seesaw and the quality of the ride. As long as the two riders on the seesaw remain equal, each can manifest his/her/its greatest potential without disturbing the balance and stability of the system. The philosophy and the physics are at one on this point.

The Tenth Principle

The Tenth Principle: Energy is neither created nor destroyed. Energy simply is. Because we know the $_7\alpha$ exists in an infinite number of forms, this principle may seem to contradict the Eighth Principle (Mass is both created and destroyed.), but such is not so. Energy is the *permanent* form of all things in all realities: Mass is impermanent. Energy is not mass or vice versa; energy has its own standards of behavior. It always is, just as it always is not. Energy is the composite reality of all planes—this earth, this universe, other universes. It's inherent in energy to "make" each of us anything we want to be, wherever and whenever. It's also possible to exist as pure energy.

Think of yourself sitting in a packed auditorium listening to a dull, boring lecture. Your physical body, grey tweed suit, pink blouse, black pumps and matching handbag are elements of your physical (mass) form. Before you dressed, your jeans and sweatshirt and the feelings, thoughts and physical activities you experienced created another physical form which disappeared when you dressed for the lecture. However, although you may change your physical appearance entirely and concentrate your thoughts and emotions in completely different areas, you're still you. You close your eyes in the too-warm lecture hall and see a statuesque blond in a flowing white gown float through a field of wild flowers high in the Himalayas. You're thirty-eight, brunette and dumpy yet there's no doubt in your mind this person is you — the energy you. Then you see the dumpy you at a P.T.A. meeting last week and then next Christmas sitting at the base of a gorgeous tree that would never fit in your split-level suburban home. Still, there's no doubt in your mind these yous are also you—just in a different form, the energy you unbounded by time, space or a specific mass form.

The $_7\alpha$ moving freely from plane to plane at will does so in its pure energy form; it can't do so as mass just as we can't transmit our mass to other realities in the dream state. Our dreaming reality, in one way, is a more real state than this one because we achieve it in the pure energy form. This is also true for our thoughts and emotions. We may also think of the $_7\alpha$ as the unit of love, the purest form of energy. Does this seem farfetched and nonscientific? Think about it. As hard as humankind has tried, love can't be destroyed. It also can't be created for it always was, is now, and forever will be. Whether or not we add the standard closing "world without end" depends upon all of us. We may destroy mass—buildings, cities, bodies—but we can't destroy energy, consciousness or love. It simply can't be done. Energy, like love, "conquers" all because it is all.

The Eleventh Principle

The Eleventh Principle: Any energy collected in forms above the level of the $_7\alpha$ isn't energy; it's an energy/mass combination. If we create energy in such a way that there is more than a photon left at the end of the reaction, or our energy collection or processing devices use more energy than they create, we know we aren't dealing with a pure $_7\alpha$ energy form. Solar "energy" and nuclear "energy" aren't really energy at all because they're composed of intermediate mass/energy states or combinations. Think of pure $_7\alpha$ energy like an egg you're going to use to make a cake. Before the shell of the egg is cracked and its contents mixed with flour, sugar, milk, the egg is an egg. When the cake is baked, where's the egg? Is the cake a "higher" form of the egg? In some ways it is, but the egg has given up its egg-ness and the cake hardly has the same characteristics as the original pure egg. Similarly the combined energy/mass states of nuclear, solar or hydro-electric energy create a cake-like conglomerate of mass and energy that doesn't express the true "energy-ness" of the $_7\alpha$.

The Twelfth Principle

The Twelfth Principle: The total energy of any state is the total of the energy of all its components at any given instant. So,

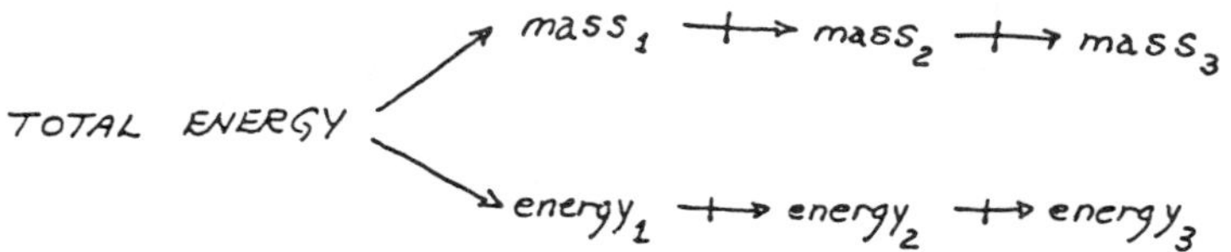

And because the term total energy also implies total mass we can also say

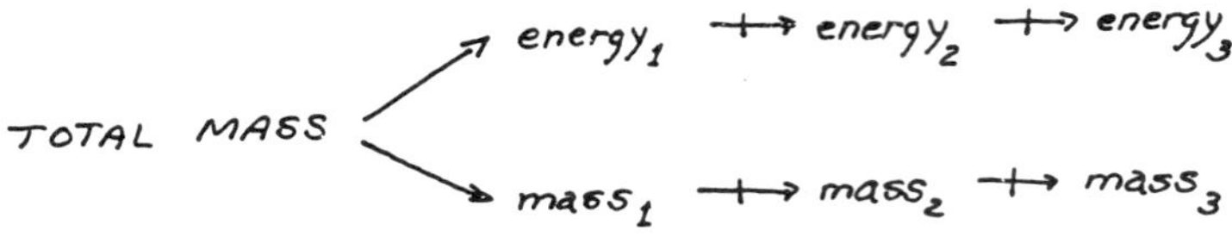

Therefore the corollary is also true: The sum of the mass of any potential energy is equal to the sum of all its component mass forms. In other words, the whole is equal to the sum of its parts.

These statements relating to mass and energy are really saying that the at-one state of any entity is the total of all energy and all mass of its component parts.

Let's examine this in terms of our seesaw with your mass form at one end and your energy form at the other. There are two ways you can look at this. First, you can see yourself in your blue bermudas, tank top and Adidas on one end balanced by all your emotions on the other, with the fulcrum being the thought process which communicates changes in one to the other. You feel uneasy. You think, "Should I have worn my red shorts instead?" You change or don't change shorts and experience the results this has on your feelings. That's the system in the here and now.

In a grander sense this balance applies to *all* physical/mass forms you ever assume and *all* energy available to you. So as you sit on your mass end of the seesaw, the infinite masses of infancy, childhood, adolescence, middle- and old-age are there with you. Similarly the energy components of these infinite forms are balancing the system at the opposite end. In such a way you are limited neither by the physical nature of that infinitely small point in time defined as now, nor by the emotions/energy inherent in that definition. You may experience the physicalness and ability of a child and the feelings of an adult or vice versa. If you always manifest both energy and mass forms equally, you'll always be balanced.

How can we be sure we're manifesting energy and mass forms equally? The answer is simple but sometimes surprisingly difficult to implement: the easiest way to be sure the physical (mass) and spiritual (energy) components of any system are balanced is to always use them maximally. If we know we aren't holding back anything, be it feelings or physical manifestations, then the system is balanced. If we know we're withholding something, storing any form of our being "just in case", we can't be balanced.

Summary

Let's summarize the 12 principles of energy:

1. If energy is created at the smallest level ($_7\alpha$), there is no waste.
2. Process is less important than the result.
3. Heat-based reactions are wasteful and inefficient.
4. Any chain reaction is unnatural.
5. Energy in one form can supply all the needs of the earth.
6. The more specific the energy-creation device, the more energy it wastes.
7. Whenever energy is stored, more energy is lost than gained.
8. Mass is both created and destroyed.
9. For every (re)action occurring in nature, some change in either mass or energy occurs.
10. Energy is neither created nor destroyed.
11. Any energy collected in forms "above" the level of the $_7\alpha$ isn't energy; it's an energy/mass combination.
12. The total energy of any state is the sum of the energy of all its components at any given instant.

Now let's examine each one of our principles in detail.

4 | The First Principle

The First Principle: If energy is created at the smallest level ($_7\alpha$), there is no waste.

The Hindu religion and philosophy has long stressed that everything in the universe is connected. These connections must then exist on the smallest as well as the largest level because the greatest interaction can only arise from changes in the least. In cases of the survival of a large number of people in a disaster such as a theatre fire, the success of the group can normally be traced to the behavior of a single individual who initiates the life-saving activities. The Jonestown mass suicide resulted from the belief of one person which was then transferred to many others. If we have people packed tightly, shoulder-to-shoulder, front-to-back, in a small room, one person's choice to scratch an ear causes motion in everyone.

We can speak, then, of every action having the ability to create change in everything else. This may be easy for us to accept in theory, but in practice in our day-to-day lives it's often hard to picture. Think of all that is as a pond of infinite size, and of every change as a pebble thrown into that pond. Regardless of the pebble's size, it eventually affects the entire pond. Although we might not notice the changes created by the smallest, we know they occur. Even pebbles so light they float on the pond's surface, affect the nature of the entire pond. In many ways the changes created by the floaters may be compared to spectators who create changes in others' direction or form but remain relatively unchanged themselves.

Natural Flow and Energy

When we speak of creating (actually filtering) energy on the $_7\alpha$

level, we speak of natural energy, that which is extracted from the natural flow of things without disturbing the natural order or rhythm. Think of an infinitely wide and long progression of people. As its participants approach (enter your reality), you ask those who like to write, draw, work with numbers or with their hands, to form separate groups. In such a way the different forms of energy can be created or filtered; no individual relinquishes identity in the process. Our people are still themselves even though some are writing, some drawing, some solving mathematical puzzles or building. None *must* stay in his or her particular group; they're there by choice and because that group is where they feel the most comfortable.

Does dividing our progression into such groups destroy the natural rhythm and order of the whole? Let's suppose the rate of flow remains constant. If we have two groups, A and B, moving from left to right

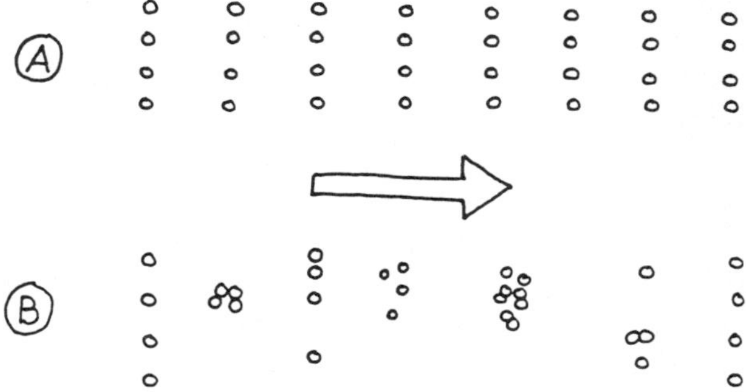

and both maintain the same rate per unit time, their flow is the same. Compare this to a marching band versus a band with majorettes and pom-pom girls pirouetting among the players. Although the girls display activity completely different from that of the band, because they maintain the same overall translation or forward motion in addition to their own unique behavior, there's no disruption.

This is an important distinction to make because it gives pure or $_7\alpha$ energy a much slower rate than current energy sources. For example, we can compare current energy form and flow to a band whose players, pom-pom girls and majorettes are uniformly distributed among the group. Each rank may have one pom-pom girl, one majorette, one clarinet player, one drummer, etc. If each row has twenty players and it takes each row a minute to pass us, we can say the band's rate of flow is twenty players per minute. On the other hand, pure $_7\alpha$ energy corresponds to a band whose pom-pom girls, majorettes, and similar instruments are grouped together. Depending on the size of the band we may only see one group of majorettes every two minutes, one of drummers every three. Although we recognize the uniform flow as faster, we can never appreciate the qualities of the various groups contributing to the total picture. Those who prefer the sound of brass or woodwinds, or thrill to the synchronized but different pace of majorettes and pom-pom girls are denied these special forms.

Pure Energy vs. Control

In addition to suffering because of its nonspecific, less-than-smallest entity form, current energy is further burdened by our beliefs regarding its storage. In essence, we may think of current energy storage techniques as the collection of all the marching and playing band members, majorettes, and pom-pom girls behind a single door which is suddenly opened, spilling them all over. Although the force created by the group present when the door opens far exceeds that of any individual, the rhythm of the music, its coherent sound and synchronized movement is totally destroyed. To be sure, there's a great deal more force and power inherent in that group the instant it bursts through the door compared to one which maintains a continuous and unobstructed pace; but because it lacks both order and flow, it's incoherent and cacaphonous and soon expends itself. Whereas an unobstructed or "unstored" band marches, performs its piece and moves on, this groups bursts through the door and creates little or no music beyond that initial burst of noise. No worthwhile music is produced, additional time and energy is required to reorganize the group; furthermore their presence blocks the paths and performance of those following, thereby insuring this nonproductive sequence will continue.

Our analogy is multi-leveled because understanding the meaning of pure smallest $_7\alpha$-level energy requires such. We may think of current energy flow—whether light, potential, heat, sound, or pressure—as marching bands trapped behind closed doors. We often believe energy is something we must *control:* turn on and off at will, jerk the door open and see the band spill out with a rush and burst of horrendous noise and fall on its face so we can slam the door and hear the bodies hit again with nowhere to go, insuring us of their presence, ready to respond to our instant demands. Current energy sources aren't energy sources at all, but rather a means for our technological society to manifest its *feelings* of impotence. What keeps the current technology of energy production alive is obviously not its efficiency, economy, or even utility. It exists the way it does because it answers a need in our timorous societal psyche to have something there on *command* and *demand* whenever we turn a switch, faucet, key. It isn't the nature of that power or even what it does that matters—it's the fact that it jumps when we say, "Jump!" We can also think of energy in terms of a prevalent view towards wealth. Most have no idea what they want money for; we merely want it. If we're poor and have no idea what we'd do with the money if we had it, chances are we'll never have it. If money is power, then its accumulation and storage becomes all that matters; this is similar to having the biggest generator in town, but no electrical appliances. It doesn't matter. Having money, power or energy is like being good: We must be good for something, for a purpose. Otherwise we're merely good . . . for nothing. These views are merely different forms of storage which we already know are self-limiting reactions. Unless the product of a reaction is continuously removed and used, the reaction cannot continue.

However, as we listen to the beautiful music of our continually marching band and appreciate the charm of the precise movements and the colors, we long to store this perfect image, sound and sensation so we can have it for ourselves anytime we want. However, we can't believe the band would *want* to respond to our needs whenever and wherever we desire, so we create abnormal, aberrant forms to give an on-demand impression. If we're wealthy enough, we buy the whole band; if we're sufficiently powerful, we take them by force; if we're intellectually creative, we duplicate them by making video or audio recordings of their activities. However, regardless of what method we use, we can never have the power inherent in the group within its natural flow by its natural choice.

Smallest Level Energy is a Reflection of the Self

By understanding how energy is currently formed, stored, and used as a function of how each individual perceives him or herself, we can see how threatening new ideas and theories can be. To say, "It isn't necessary to turn a switch to get energy," isn't offering a particularly reassuring solution to listeners who seek the reaffirmation of their own power inherent in that on/off switch. To them the form of the energy is immaterial. It's the ability to turn it on or off, to control its behavior, to *make* it do what they want it to do when they want it done—*that's* the reason and justification for energy in that form, regardless how wasteful, inefficient or even dangerous its production and storage might be.

Extending this line of thought into the nuclear realm, what is the real lure of nuclear energy and weaponry? It's not destruction, *per se*, but controlled destruction when and where we want it. The technological human race doesn't want power to create and destroy: It wants power to create and destroy *on its own terms*.

Right now most of us clamor for clean energy, but suppose energy is simply a different word for love. If you doubt the practicality of this possibility, let's substitute the word "love" for energy whenever it occurs in a few of our principles:

- Principle One: If love is created at the smallest level there is no waste.
- Principle Five: Love in one form can supply all the needs of the earth.
- Principle Six: The more specific the love-creation device ("If you love me you'll do _____, _____, _____.") the more love it wastes.
- Principle Seven: Whenever love is stored, more love is lost than gained.

- Principle Nine: For every (re)action in nature, some change in either matter (body) or love occurs.
- Principle Ten: Love is neither created nor destroyed.
- Principle Eleven: Any love created in forms "above" the level of $_7\alpha$ (the smallest level) isn't love, it's a combination of love and something else.
- Principle Twelve: The total love of any state is the sum of the love of all its components at any given instant.

We're sure you can think of many other examples where the two words are interchangeable:

"Love/energy makes the world go 'round."

"Love/energy conquers all."

"God is love/energy."

Remember: Most of our current views of energy and love are in terms of control. Whether we speak of lighting a 60 watt bulb, detonating a nuclear weapon or proposing marriage, form, utility, efficiency, safety and economy invariably take a back seat to control. Having energy, like having money, power or love, for many isn't the issue; it's the need to control it that creates the problems. It wasn't money that was/is the root of all evil—it's the *love* of money that creates the problems. Perhaps the arms and strategic weapons "control" we so desparately seek isn't putting down our weapons or beating our swords into plowshares, but the freedom to embrace one another. Combined forms such as arms control committees, and delegations require a lot of time, energy and money yet produce very little in the way of lasting results.

The $_7\alpha$ and Nuclear Energy

In our discussion of the value of creating energy at the smallest level, we must consider the current status of and argument for nuclear energy. At one time the nucleus, or more properly the nucleons (neutrons and protons), were considered the smallest entity and look what happened: Many smaller entities were discovered and more appear every day. How can we be sure this time that the $_7\alpha$ is the smallest entity? If the $_7\alpha$ is the smallest entity and disruption of larger moieties (neutrons and protons) has proven so detrimental, how can we be sure energy collection on the $_7\alpha$ level won't be worse? That answer lies not in elegant theory of revolutionary devices but within each individual. To be sure, the physics and mathematics make a most appealing case for the $_7\alpha$ being the smallest unit and its energy the cleanest and safest, but even those who already know and understand the theory and its probable validity must deal with their own fears.

Obviously whenever we're presented with a new concept of energy and existence that requires change, even in the most mundane, mechanical aspects of life, we must recognize ourselves as capable of *conscious* change if we're not to feel threatened.* If we don't first see ourselves as capable of making infinite change by choice, it matters little what we're given; we won't make even the most rudimentary adjustments to accommodate the gift. To the vast majority of the technological population, light is something one has always turned on and off with a switch. Electricity goes through wires and comes from generators or batteries; cars have piston engines that utilize gasoline or diesel fuel. That's the way it's been for years. Energy choices have been negligible and more often left to the scientist, the inventor and the entrepreneur.

How many *conscious* choices regarding energy production have you made in your life? We're willing to bet the answer is, "Not many." Although we acknowledge the choice among numerous devices to use energy, we tend to think of energy as something that always was, but is running out.

Perform the thought experiments comparing energy with love, using analogies that make sense and are familiar to you. It's a perfect analogy in terms of rotational physics and philosophy, but it's often not representative of current scientific or philosophical thought. For example, people who have false beliefs about energy tend to carry those same beliefs about love, and vice versa.

Many who now fear nuclear power, radiation, waste, and pollution also fear love. "If I totally love John, what happens if he

---

*For a method of producing conscious change, see *The Power of Choice* by R. Fritz and B. R. Smith (1982, Fainshaw Press).

leaves me (doesn't feel the same way, finds someone else, dies)? I'd better hold back a little, just in case—so I won't get hurt, so I'll always be in control." So Mary holds back in one way or another because she's fearful and sure enough her fears, in one way or another, are fulfilled. "I wasted the best years of my life on him," says Mary bitterly. What remains of the less-than-total love, the impure form that started the reaction, has all the negative value and ramifications of a pile of radioactive nuclear waste.

Not surprisingly those who find even a conscious recognition of the crucial role love plays in their physiological, psychological, and psychic well-being threatening often displace their energy-love awareness one step further: They create a god or gods. There's absolutely nothing wrong with a god-concept as long as we realize:

- We create and co-create gods for our own purposes. As such we can co-create nice, loving gods (the Christian God) and some real tough guys (Jehovah, Allah).
- Gods exist at *all* levels.
- We all possess the greatest power of any god we can imagine—free will and choice.
- There is a truly multidimensional God which is the sum total of all that is.

None of this discussion should be construed as an attack on any religion or, for that matter, any individual's feelings about love and/or God. If you're comfortable with your own beliefs, you'll feel no need either to defend or compromise them, and what we say can in no way be threatening, even if you disagree. However if you feel yourself getting a little hot under the collar, you may want to explore your beliefs regarding energy, love and God(s) a bit further. Whenever we give someone or something power over us, it's incongruous that we would simultaneously feel we must protect and defend the very thing or being to whom we delegate that authority.

We introduced love and god-concept analogies to energy because an understanding of how energy is created requires we recognize

- We can change, and the only consequences (or sequences)

are those we define for ourselves;
- Energy, its formation, collection, and use is inherent in each one of us.

For some, the scientific explanation of $_7\alpha$ energy formation is sufficient. Others may prefer the less classical, more romantic approach of philosophy or theology, while still others need both. For some nothing will change their beliefs.

To see how these different approaches can help us understand the value of energy creation at its smallest level, let's take the famous equation, $e = mc^2$ and express it mathematically as

$$\frac{e}{m} = c^2$$

This can also be expressed in some of the following ways:

$$\frac{love}{wisdom} = enlightenment \qquad \text{(theologically)}$$

$$\frac{mind}{body} = spirit \qquad \text{(philosophically)}$$

and in rotational mathematics as $\quad \dfrac{0}{1} = \infty$

Although it may appear we are highly critical of current energy production and storage systems, we do believe the general evolution has been in the right direction. We have successively gone from larger to smaller entities in our search for proper energy collection. If we look at the progression we see

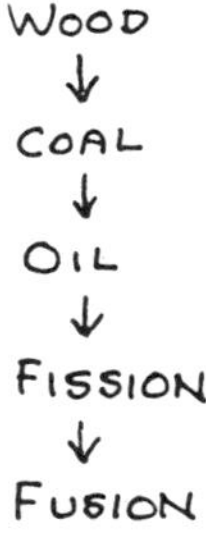

each time dealing with smaller entities delivering more heat per unit volume.

Because the direction has been proper and benefits realized, it shouldn't require too great a mental leap to realize that if we deal directly with *the* smallest entity, our problems of inefficiency and waste should disappear and the benefits increase proportionately.

5 | The Second Principle

The Second Principle: Process is less important than the result.

Before we discuss some of the scientific ramifications of the Second Principle, let's look at some of its philosophical implications. For years managers in both public and private sectors have been trained in the use of management by objective (MBO), where the word "objective" is synonymous with "goal" or "result". However many organizations haven't had much success with MBO because they focus on the process, the "how", rather than the result, the "what". Does focusing on process only result in more process? Look around you: How many times have you run across people who insist a certain procedure *must* be followed to the letter or some form fully completed when these activities seem to have little significance relative to the accomplishment of the stated goal? We've all heard the story (fictitious, we hope) of the patient who expires in the emergency room while the admitting clerk collects information about insurance coverage, age and occupation of all known relatives, and so on.

Unfortunately, those who recognize this misplaced emphasis often set about to destroy the process. Bucking the system—going up the down staircase, sneaking in the exit—is merely a different form of process and often involves more process than that we seek to destroy. Those who create process and those who constantly look for ways to violate that process are equally process-dependent.

Anyone who's been involved with process, concentrating on the means rather than the ends, invariably realizes that the creation of more process eventually becomes the object of the process. At this point the more observant realize the relative uselessness of such a system. Regardless how many times we answer the questions of how and why, we must still deal with the result—which has often become something quite different during the evaluation of the process and necessitates starting all over again. Those who realize they're caught in a nonproductive cycle may respond in one of three ways:

- Shift their emphasis to the result.
- Become overwhelmed with the uselessness of their work and sink into depression and inactivity.
- Become determined to work through all the process *and* know the result, making them "short on time and long on jobs." This can lead to impatience and frustration.

For example, Dave Duryea's stated purpose at Ho-hum Electronics is to thoroughly inspect one hundred digital pulse monitors daily. Because Dave is keenly interested in electronics and spends much of his free time expanding his knowledge about this field, he learns to perform this task well in less and less time. Because he stays up late to attend classes or read material that increases his job performance, he sees no reason why he must be at work at 8:00 each morning. Were Ho-hum goal oriented, if they wanted the most competent inspectors monitoring their product, they could accommodate Dave's approach to his job. However, Ho-hum is process oriented: It's more important that their employees arrive at 8:00 a.m. and leave at 5:00 p.m. than that they're happy and do their work well. Eventually Dave and his supervisor become embroiled in a battle over this deviation in process. This affects Dave's attitude, then his work, and he eventually quits or is fired. His replacement is neither as knowledgeable nor as conscientious about his work, but he is punctual and Dave's supervisor heaves a sigh of relief.

Therefore, negative emotions—depression, impatience, frustration—are commonly produced as a result of process. No wonder the phenomenon of job burnout is so prevalent: Most of these folks are so deeply involved in process, many aren't even capable of knowing what the result is supposed to be.

So if negative emotions are created by process, there's often no way we can get beyond them to the result:

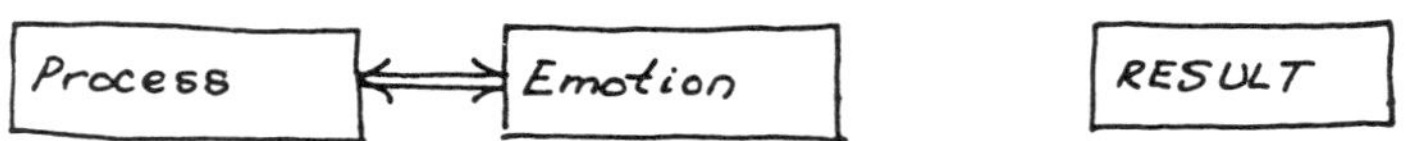

Therefore, we can say that process is emotion-dependent. Those who concentrate on the process—the hours, the hows and whys— as a sequence of predictable events, are governed by negative emotions; those who concentrate on the results are governed by energy, a sense of accomplishment, love or other positive emotions.

During the next few days, pay particular attention to how much time you spend getting from "here to there"—picking the right time or the right sequence of events to get from a physical here to a physical there, a particular process for communicating with another ("I must be sure not to call Helen when her favorite show's on." "Bob hates to be bothered at the office."), or the need to define the way you feel while experiencing certain events. (Do you feel guilty if film clips of starving millions don't bother you? Do you expect your children to react to their presents from you a specific way? Do you feel hurt or cheated if they don't?)

Habits and Facilitated Responses

To understand process we need to know the difference between habit and something called the facilitated response. A facilitated response enables a lumberjack to sense a slight veering of a falling tree and instantly jump two meters instead of his planned one meter to avoid being crushed. The term refers to an event more than a process. The lumberjack usually does one thing; this time, based on his knowledge, experience and intuition, he does something else.

Compare this to habit which exists by virtue of process. Where the facilitated response is a result more of preservation instincts and a wide range of knowledge, the habit or pattern arises out of repetition and boredom. Even so-called good habits often result because individuals believe there's nothing better to do. If our logger downs one thousand trees, always moves one meter and is always safe, he could easily be lulled into the belief that it's unnecessary to pay attention to his goal, the downed tree *and* his safety. Consequently when circumstances require a two meter jump, he fails to respond. In essence, the process and figures (which we're told never lie) say a one meter jump is adequate and that becomes incorporated into his pattern. Because he pays more attention to the established process than the desired results, he makes an unnecessary and fatal error.

Breaking Away From Process Orientation

Can we break out of the trap of process, patterning, and habit? We can, but first we must recognize the process itself; then we may apply one of two methods to change our process orientation to result orientation. How do you know if you're involved in process? Ask yourself a simple question: What part does my work play in the overall product or goals of myself and this organization (family, committee, team)? It's surprising how many workers have no idea what happens to the part or paperwork they create once it leaves their work station. For example, Janie Seymour was hired to operate a punch press at United Car Manufacturers and given a production goal of 4,000 punched parts a day. Once she mastered the machine, Janie easily maintained her goal and soon became quite bored. Fortunately her supervisor recognizes how limited the job with its current definition is and takes Janie on an extensive tour of the plant. She sees where the raw materials enter and the steps preceding her task. Then she follows her part as it's incorporated into the more complex automotive inner workings. After her tour Janie realizes those parts she makes are vital, not only to the performance of every United vehicle manufactured at that plant, but also to the safety of every driver and passenger in those vehicles. Because she understands how her part fits into United's overall goal of creating quality cars, Janie now has her own goal: She wants to make the best punched parts ever made by United or any other company. If we're doing something and we're not too sure why, chances are we're involved in process.

Once we recognize we're involved in process, we can use different methods to become more result oriented. One method involves merely changing one step in the sequence. This is the primary function of many medications; they interrupt some routine or pattern in the disease-producing organism, giving the body time to recover. While the medication is altering those micropatterns, it also disrupts any pattern the person may have established regarding the disease and his or her ability to fight it. Often just the act of taking or doing something different is sufficient to break the non-productive, process-oriented cycle of depression and self-pity which rarely acknowledges as its goal a healthy state. ("This is the same bug Carla has and I know I'm going to be miserable for two weeks.") Just plunking down good money for a professional opinion and/or medication is often all it takes to make us more goal-oriented: "Dr. Smithers said these new pills would have me up and about in no time and he's right!"

Think of yourself caught in a traffic jam on your way to visit Aunt Harriett whom you dearly want to see. You can stay fixed in traffic and possibly become angry and frustrated (a process producing negative emotions) or you can focus on your goal of seeing Aunt Harriett. The latter approach has two probable effects:

- The negative emotions are eliminated or at least reduced.
- You make some physical change—change to a different road, stop for a while and rest.

Changes in process never work when the goal is simply to change the process. Unless the result is the goal ("I want to get to Aunt Harriett's before noon." "I'm so happy I'll be seeing Aunt Harriett again.") rather than the process ("I am sick and tired of driving in a pack of jerks."), any response is most short-lived.

The second way to overcome patterning or process is via a sequence of consistent minor changes. Isn't this just the same as creating a new process? It is if you concentrate on the changes themselves and not the result. For example, physicians often recommend dietary changes to break established detrimental patterns. Patients then respond to these recommendations in one of three ways:

- They implement them, expecting the *changes* alone to "cure" them.
- They consider the changes a violation of their freedom, a threat to their established way of doing things, and reject the advice.
- They accept the recommendations, make the changes and then ignore them: The changes are neither good nor bad, just part of what becomes their normal behavior.

If they believe the changes are the same as the cure (eating a low-salt or low-cholesterol diet will repair the damage done to the heart and circulatory system by years of all sorts of unhealthy practices), they soon become disenchanted with this bland diet that doesn't make them young and vigorous again. If they consider the recommendations a violation of their freedom, the process is obviously more important than the goal. Why does Harry keep smoking even though he knows it aggravates his asthma tremendously? Because Harry is more afraid to change *now* than he is to die sometime in the future. This common state is a paradox because the belief that the refusal to change violates one's freedom is most obvious evidence as to how much one is enslaved to a particular process. Those who accept recommendations, incorporate them into their routine and then ignore them are much more likely to experience a cure. Think of all the dieters who bounce from one fad diet or exercise program to another without ever experiencing much benefit. Their conversation is invariably dominated by references to the mechanics of these programs, not their plans for their new slimmer selves. Those who have such goals invariably achieve lasting results much faster and seldom attribute them to any specific formula diet or exercise program. When Janie Seymour sees her goal as the production of a quality part in a quality car rather than eight daily hours of boring, repetitious work, everyone benefits—Janie, her employer, consumers, stockholders and suppliers.

Therefore, a change in process has to be made to change the results. The process is definitely dependent on the results and vice versa. But if the results are merely changes in the process rather than a point beyond the process, a point that further enhances one's ability to *ignore* the process, they're not lasting. If Janie's boss merely ups her quota from 4,000 to 8,000 parts a day to challenge her, it may alleviate her boredom for a while but she'll soon be back in the same rut again. Similarly the person who loses weight and keeps it off, quits smoking or drinking and remains in that state, does it without thinking because they see a goal beyond the process. The process of losing, quitting, changing isn't their goal; they focus on a result beyond the process that may or may not be related to the process at all.

Therefore if the changes are first viewed as good and desirable and then ignored, even the most rigid, long-standing patterns can be broken.

Process Orientation in Physics

Not surprisingly, process rather than result orientation creates similar negative effects in physics as in business and physiology. Just as process orientation produces a rigidity or stress within the individual, so it creates rigidity and stress within the physical world. Think of process as the bridge between an existing condition and a desired change or result. How process affects the system is identical to the effects produced by stress and its resultant tensile and compressive forces on the bridge. Consequently the average engineer, including energy engineers, are trained specifically in process—the best method of getting from here to there, of creating the wherewithal to get from here to there. It's an orientation thoroughly built into present science—deductive reasoning, the "scientific method" creating reproducible results based on intricate process; and it pervades all levels and forms of education.

Within energy creation the paramount *goal* is invariably to produce more electricity faster, cheaper, and with less waste. However, the total impetus of the system centers around changes in *process:* new forms of wiring and transformers, new forms of collectors and photo-voltaic cells. However, the result is invariably the same—stored electricity.

We need only look at our easy embrace and acceptance of wood heat during the "energy crisis" of the seventies and the continuing allure of the visible fire to illustrate

- The willingness of a large number of people to use different forms of energy and accept the changes those forms create in their lifestyles and personal philosophy.
- People's intuitive recognition of different forms of energy (results) being of more merit than process.

In other words, people are willing to accept different results (wood vs. electrical energy) as well as changes in process to achieve the same results (different ways of producing and storing electricity).

We briefly discussed the role of stress in process orientation, physiologically and psychologically; now let's see what happens in energy creation. If we take our basic fire and expand it, putting it into Franklin's stove and eventually into a reactor, we can see how these structures are all processes of energy production. The caveman discovered fire and created heat to cook his food and keep himself warm. Furthermore, he recognized from his neighbor's envy that fire was power and therefore began designing ways to store it, ways to preserve and carry these glowing coals instead of relying on lightning, flint, or two pieces of almost anything to rub together to create friction and heat. Any soldier, boy or girl scout with the least amount of survival training knows creating fire isn't that difficult and at most one must endure 12–16 hours without it if conditions are extremely wet. However the belief that fire, like love, keeps away wild animals and "wild" bacteria and other such evils in our food is so strong, the idea of not having matches in a survival kit is seen as the worst thing that can possibly happen.

Creating a New Philosophy

If our result is energy which is clean, safe, cheap, and independent of earth's resources, then we need a new bridge or process to bring us to that destination. How are bridges built to replace existing structures usually constructed? In general, construction of the new occurs simultaneously with the use of the old. Those who continue using the old see the new bridge every day and make judgments regarding any differences they perceive. They note not only the bridge itself, but where it ends on the other side of the river. Let's assume for our example that the new bridge leads directly into a city whereas the old structure led to a dirt road, which, via uncomfortable and even dangerous meanderings, wound its way to the city by a roundabout way.

The people observing the new bridge note not only the bridge itself but where it ends on the other side of the river. Pretty soon they notice that, rather than ending out in the boondocks where they must seek out other forms of transportation into the city, the new bridge takes them directly to the city where they can easily walk to whatever they need. Eventually when the bridge is complete they begin to use it, some more readily than others. However the old bridge isn't destroyed nor is it decried by the creators of the new; in fact, they simply ignore it. Similarly those who are initially caught up in the unique construction of the new bridge soon ignore this aspect in favor of all the opportunities available in the now easily-accessible city. In a brief period of time they change their entire lifestyle. Because the city is now so close and the bridge so safe and convenient, they needn't worry about storing supplies. They're also free from the fear that the new bridge itself will collapse.

Meanwhile fewer and fewer choose to use the old bridge to the old wilderness. To be sure the bridge-keepers make a few half-hearted attempts to maintain it and add to it, but they have been offered enticing jobs in the city and feel little desire to maintain something so mundane and inferior as the old bridge. In a very short time with no threats or angry words, no destructive behavior, the entire view of the population has changed. Where once their lives were dominated by a process yielding the most primitive of results, they now ignore both the old and the new process and concentrate on the result, creating even more results. No one is denied access to the old process; no one is forced to accept the new. *Each individual* accepts or rejects the processes and results *by choice:* That is the *only* way any lasting change can be accomplished.

In our bridge analogy we speak of building a new structure which others can choose to

- Work on or not work on,
- Use or not use.

Although the sequence of events seems reasonable, we are left with the old bridge which we implied would soon collapse from disuse and lack of maintenance. Although this would assure the presence of our new bridge and new lifestyle ("You can't go home again."), we can't deny the wastefulness of such a procedure either. Isn't there some way we can use the first bridge to create the second? To go from what we know and understand to what we don't?

There is but it must be done very carefully. Building a new bridge is one thing; dismantling one bridge to build another at the same time is quite something else. As long as the old system remains in an unsafe condition, as long as no new bridge is constructed, there is no chance for improvement. However, removing the old bridge may bring out fears and concerns: "I liked the old bridge; I was used to it." "I have my money tied up in the old bridge." "Shouldn't we hang onto the old bridge in case the new bridge fails?"

Old beliefs, especially those grounded in process, die hard. Although we may argue familiarity breeds contempt, it also creates a certain amount of security, too. We need only look at the success rate of incumbent politicians versus newcomers to see how this thinking works. As one voter noted, "Sure I know he's a jerk, but the other guy could be worse." Better the known jerk than the unknown fool; better a known process even if inefficient, unreliable and even dangerous, than an unknown one. Whether contemplating a new bridge, a new form of reactor, or a physiological change, the fact remains that some very intelligent people will create some very complex rationale to defend the existing process and this can only be dismantled with the greatest amount of care and delicacy.

Changing Process in Energy Production

Can we dismantle (change) the process resulting in nuclear reactor energy production and replace it with a passive, safe energy creation process? Can we maintain our *unsafe* old bridge in a *safe*, passable manner as we gradually dismantle it to build a new one? This can certainly be done; however we must first learn to look at our old process a new way.

What does this mean relative to nuclear energy? A great deal. For example, scientists and engineers believe the process of nuclear fusion—the "building up" of helium isotopes from those of hydrogen—is the epitome of energy production. Hydrogen-helium fusion is indeed important, but placed as it is within the unstable process of current nuclear energy production and storage, it's in a most precarious position indeed. Artificial fusion requires extremely high temperatures, even higher than that of the sun because, theory says, there can be no fusion without heat. Thus the current process of fusion requires tremendous amounts of heat to produce heat. In theory that heat is then converted to electricity. In fact, so much heat is produced in both the process of nuclear fusion and/or fission, only a small portion can be converted to electricity. The rest becomes waste. (Remember those gigantic cooling towers at Three Mile Island?)

However, suppose we look at the natural fusion occurring all around us as an energy *source* rather than a *process* to be sped up or duplicated unnaturally. From this point of view we can recognize natural fusion products such as sunlight and cosmic rays as abundant sources of naturally fissionable or polarizable products. For example, our photon with its two perpendicularly-bisecting 50/50 $_7\alpha$ is a natural fusion product

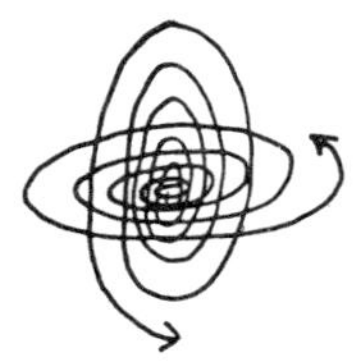

which we can then separate (fission),

into horizontal and vertical components. If we then allow these components to stretch out into the 0/100 state or close to it, we have

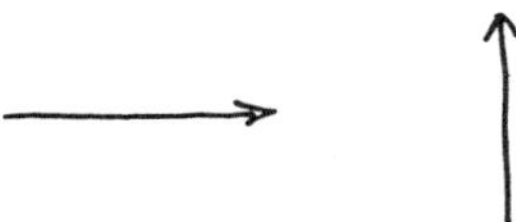

and then by recombining or fusing them

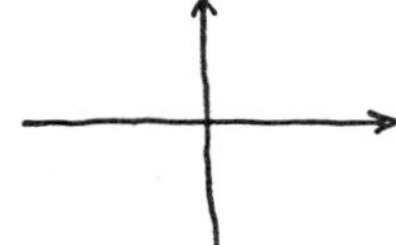

we can create pure potential.

By shifting our emphasis on fusion as an artificial process within other artificial processes to fusion as a natural energy source, we can see it in a new light. Viewing fusion this way enables us to see energy production as the separation of natural energy forms rather than the breakdown of mass with its attendant inefficiency and dangerous, intermediate wastes.

Doesn't it make sense that the most efficient processes of energy production would take what's naturally available and filter or separate it into the particular components necessary to fill our needs at a particular time? In other words, rather than using a process to break down and form new mass to create energy, why not use a process whereby existing energy is filtered and realigned to produce a purer, more efficient product?

Think of a thousand people milling about in a large hall. Current fission techniques are similar to rolling bowling balls into the hall hoping to knock the 50 physicists in the group through the walls so they may be collected. Current fusion techniques would heat up the walls, driving the entire group towards the center, again hoping the physicists would find each other. Rotational physics would place a series of openings throughout the room designed to attract various individuals, thereby separating the entire group. In addition to the physicists, the process would identify and collect the teachers, mechanics, clerks and lawyers.

Although fission and fusion are viewed as processes *creating* energy from mass, both actually do use a type of filtering mechanism. However rather than using a highly elastic substance which can expand and contract its "pores" with relative ease, fission relies on the highly dense, brittle structure of uranium and plutonium.

Compare the course of a ball passing through the aligned openings in Group A

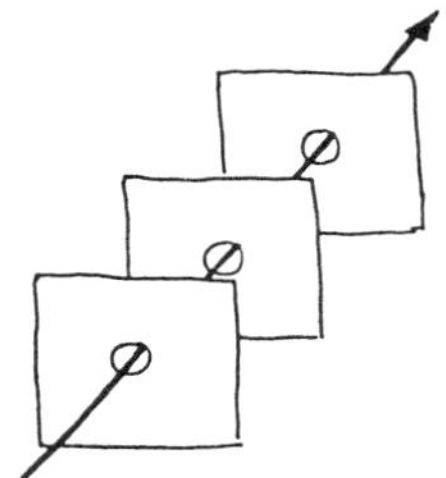

versus Group B:

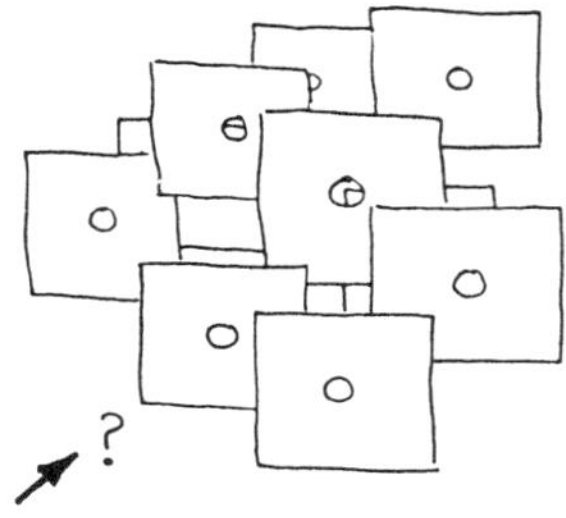

Although both A and B have the correct size openings to allow passage of the ball, it takes the ball much longer to traverse B's more complex system. Furthermore, because the ball interacts with the B layers more, it produces more heat due to friction which is essentially wasted and lost. Because we want pure potential (the form in which the ball enters the maze) rather than heat, each friction-producing interaction decreases the amount of pure potential that will eventually be produced.

Basically what happens in a fission reaction is that a great deal of energy is used to blow a complex lattice apart into more manageable units thereby permitting the flow of only certain energy forms. However because fission is at best a random process, both the resultant filters and their filtrate or product are impure. The result is incoherent heat and waste, and "used" filters now so structurally altered as to be unnatural and radioactive.

Current theory maintains the process whereby energy is released in fission involves the breaking of intrastructural bonds while the fusion process creates energy because the bonding inherent in the larger nucleus is less than that inherent in its two component parts. Rotational physics maintains the substances in both processes merely function as larger or smaller semi-permeable membranes or filters for the passage of $_7\alpha$. The fact that the mass form *changes* is evidence that the energy collection process is incomplete and inefficient. As any cook knows, you expect a sieve to separate lumps from the gravy. If the gravy and/or its lumps destroy the sieve, you might consider re-evaluating your gravy, its lumps and the process. If the sieve does what you expect it to do, is unchanged, safe and reusable, you wouldn't hesitate to use it again.

When we speak of the process of energy conversion—mass to energy—all mass converted is converted to energy. This doesn't mean the total starting mass *must* be converted; it does mean that whatever is converted is converted *totally* and whatever mass remains is *unchanged*. Compare this to a dream state wherein part of you is in a completely different (energy) form and there's "less" of the total you here; however, that part of you remaining is complete and stable in its own way. It's not waste. Or think of eating a "self-healing" apple—every time you take a bite, the apple heals itself and looks unchanged. Depending on the process used, the apple can decrease in size with each bite (conversion) or always remain the same size.

If fission and fusion are primarily energy filtering rather than creation processes, the goal or desired result should be the *collection* of energy. However the current emphasis on process is so great, it's as though we're looking at an insect-clogged screen or dirty filter as the *source* of fresh air or clean water, rather than a mechanism to refine and change it. At this time the filter's become so gummed up as to be less pure than that which it's supposed to filter. Such is the nature of becoming bogged down in process.

Current systems could be run at 100% efficiency or even higher if the elements (uranium, plutonium, hydrogen) were removed from the systems and replaced with air or water. Substances presently being used require so much energy to make them even inefficient filters, the whole process becomes an abomination. Both fission and fusion reactions are so unstable, unsafe, wasteful, and destructive they have "utility" only in one area—weaponry—where no control and total destruction are the goals. Neither process as it now exists has value as an energy-producer for non-destructive purposes.

Even now, although supposedly non-destructive energy is being produced via fission, each reactor is armed with its radioactive weaponry in the form of waste, continually attacking the environment. Is it any wonder there's such passionate emotion directed against these reactors? Is it any wonder why such systems are becoming more difficult to build and maintain?

If we put more energy into these systems than we get out, why do we use them? We do it so we can produce a form of energy we can *store*. That is our real goal; not safe, efficient energy but energy we can store so we can have it on demand. We need only examine how a *decrease* in something like savings affects our entire outlook. We stop doing things we want to do, not because we can't afford them now, but because it could mean we won't be able to afford to do something else in the future. As long as we believe we must store, such wasteful systems will result because storage *is* waste. Any system which produces a surplus is inefficient; it doesn't produce something better—it produces waste. Anything beyond what we need right now isn't gravy; it's garbage, excess baggage, that must be carried around, worried about, fussed over.

Does all this mean humankind is meant to live a most spartan life day-to-day, hand-to-mouth? Indeed not. It means we're meant to lead the most full, fruitful, luxurious life *now*, rather than in our dreams of the future. Every time we do without today, we establish the process and pattern for the future. It doesn't work the other way around.

In this chapter on process and result we touched on the nature of heat-based reactions. In the next chapter we'll see how that heat makes a wasteful process even more so.

6 | The Third Principle

The Third Principle: Heat-based reactions are wasteful and inefficient.

In this chapter we're going to talk about reactions in which heat is produced as a function of the breakdown or decay of mass. Because all of the heat isn't used, such reactions are wasteful and inefficient. Even though some may say the purpose of heat-based reactor systems *is* to produce heat, because little of that heat is used as heat or any other energy form, it becomes waste.

One of the major deterrents in the development of full-scale fusion systems for energy production is the need for tremendous amounts of heat to initiate these reactions. Whereas fission reactions create their own initiating subatomic "heat" via radioactivity and its resultant instability, the necessary heat to start a fusion reaction must be supplied externally. It's the difference between a lot of kids fooling around in a big box until it eventually starts moving versus someone having to force the kids into the box first before it can move. However, if we calculate all the energy that goes into converting uranium ore to its final form as fuel, we can see both systems require approximately the same amount of "heat" or energy to overcome the inertia created by such unnatural processes.

Heat Production, Inertia and the Streaming Effect

Because wasteful heat-producing reactions and inertia often go hand-in-hand, let's review the concept of inertia in terms of Newtonian physics. Essentially Sir Isaac proposed that a body or system at rest tends to stay at rest unless acted upon by an outside force. The magnitude of that force is usually referred to as that necessary to overcome inertia. To get an idea how this works, place a reasonably heavy book on a table and press your finger against it:

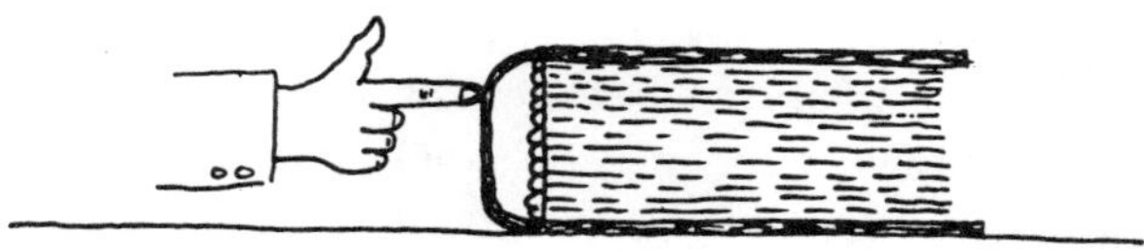

Now begin applying pressure to the book; notice it takes more force initially to get the book moving than it does to keep it sliding along once it's in motion. Or think of going sled-riding, skating or riding a bike. At first it takes a lot of energy (force) to get going, but once you're moving it's easy to keep going—sometimes *too* easy!

Inertia is the physicist's main nemesis when it comes to energy reactions. As soon as we must add a large amount of energy in any form to overcome inertia, the system is inefficient. We wind up losing at least some of our *result* in the *process*—and we must also contend with the effect of this external energy on the rest of the reaction. This is like inviting the most uncouth bore to serve as doorman at an elegant dinner party. Once all the refined guests are inside, they must still deal with this ill-mannered addition to their group even though his job is done.

Can we avoid such counterproductive approaches? Sure; if we're willing to change the way we look at things. The goal in clean energy production is to take advantage of the streaming effect: Once a single photon or other pure energy form enters a reaction, others follow. How often is it the courage of one individual voicing his or her convictions that stirs an entire group to action? Compare the group's response to input of one of their own versus some unknown even more numerous or powerful group exerting force on them from outside. More often than not, such outside force leads to resistance. Thus the most efficient way to overcome inertia is to supply the smallest quantum of that which we wish to produce. A water pump need only be primed *once,* it then can pump unassisted from then on.

Heat-based Philosophy

So true energy production requires only small amounts of energy to initiate its reactions. Notice how we speak of supplying the smallest $(_7\alpha)$ quantum of the energy form we wish to *produce*, not that which we wish to *alter*. If we want energy in the form of motion, we provide motion; if we want heat, we supply heat. In other words, we create a circular, continuous, even-paced reaction rather than a linear, geometric one like nuclear fission. Essentially, we're talking about creating dull, tortoise-paced energy production for a society geared to their hare's desire for immediate on-demand energy gratification. This isn't to say our hare-nature is ill-advised, but rather that there are many, more productive outlets for it than the creation of erratic bursts of energy that not only require more energy than they produce, but also produce more than we have the capacity to fully use at one time.

Because heat-based reactions are already acknowledged to be wasteful and inefficient, the fact that we continue using them and even develop greater reactions based on this premise indicates philosophical as well as technological confusion. If there's some flawed or erroneous philosophy afoot, we ought to be able to spot it in other areas. For example, let's look at some heat-based reactions in business. If an organization is heat-based as evidenced by the presence of anger, fear and frustration, what's its most likely product? Exactly: more anger, fear, and frustration. We get out what we put in; we reap what we sow. Organizations who seek to serve their customers must first serve on the smallest level; the least employee must be treated as the greatest, most respected client. Those who produce a product must first produce or oversee the production of that product's smallest part so *it* is perfect. Once the smallest unit is perfect, the rest of the product falls naturally into place. It's the conversion and utilization of the first quantum of energy that justifies the system, not the last. It's the amount of energy required to move a rock the first millimeter that's the measure of output, not the energy achieved by the rock at its point of maximum velocity. If company president Carson Updike keeps his assembly line functioning by constantly threatening his workers ("If we don't get this shipment out on time, you're fired!" "There's plenty of others who'd like your job, Caruthers."), he can expect less-than-perfect products. Regardless how concerned Updike is with his customers' needs, chances are he'll be offering an inferior product, one that reflects his relationship with his lowest level employees.

So often we view overcoming inertia as overpowering the entire mass when in reality we need only convert one quantum of it with an equal amount of energy. The root tip creates tremendous power as it bores through the earth, but it doesn't overpower all the soil around it. A few cells in the tip interact with a few particles of soil which then give way. In such a way "streams" of root hundreds of feet long easily find their way through seemingly impenetrable surfaces.

Overcoming inertia in rotational physics merely means gently opening the door to our lattice mechanism. However, in order to open any door we must understand how it works and in which direction it opens. Imagine banging, pounding, incessantly ramming a door that actually opens *toward* you rather than away from you. You may be able to force your way in, but you may also damage the door and what's immediately beyond it. It might make you wonder why you went to all the trouble—"I'm exhausted and there's nothing here but a bunch of rubble." Had you opened the door toward you, you might have discovered a most exquisite display awaiting your inspection. So how we open the door, how we initiate any reaction, determines what we have in the end.

Energy Collection, Fission and Fusion

Anything requiring more than the smallest amount of mass, energy, or any combination to initiate any collection reaction is wasteful. As collection rather than energy-creation mechanisms, both nuclear fission and fusion fail miserably, consuming far more energy than they produce, producing a form less pure and useful than that with which they began. Regardless how reasonable it seems that heat has no place in a reaction whose desired product is electricity or potential, what can we use to initiate the reaction if we eliminate heat? To answer this, let's look at fission and fusion in terms of what they do—their results rather than their processes. Fusion collects or "creates" energy via contraction or compression, fission via expansion or separation. Regardless what substances we place in a reactor and compress, any incoherent energy passing through the system is separated in one form or another. Similarly, if a substance is expanded any incoherent energy is filtered in a completely different way. In both cases the form created is different from that of the original substance(s). That the forms *are* different is one of the dilemmas faced by those involved in developing fusion systems. Most are *intuitively* aware the energy created via fusion won't be the same as that resulting from fission, but they don't understand how or why the two will differ. From our brief discussion of lattice mechanics it's obvious that fusion, by virtue of its compression, produces much smaller "pores" and therefore permits passage of only the more highly linear or compact spin moieties:

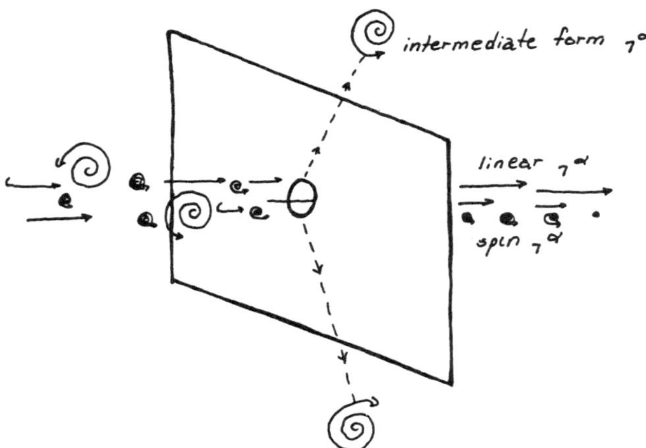

Think of fission and fusion lattices like metal strainers and paper coffee filters. Both permit the passage of liquid, but the paper filter produces a purer form than the strainer. If we place a tablespoon of ground coffee in the sieve and the filter and pour boiling water through both, we discover many more grounds in the beverage produced by the former than the latter. However, suppose coffee containing many grounds was all we'd ever known. To us, it was the ideal beverage and we designed systems to create more of it faster. That being the case, how would we view that cup of filtered, groundless coffee? Would we say it was better or worse? Would we even recognize it as coffee at all?

What if what comes out of the fusion lattice is a form of energy so alien to either heat or electricity as we now recognize them that using it would be like trying to maintain a system on apple *seeds* which was designed to use the entire *tree*? Although we may argue using the seed form is infinitely more efficient, the fact remains it's virtually useless to us in that state. By the time we alter it sufficiently to fulfill our defined needs (plant the seeds, nurture them, wait for them to become trees) we use far, far more energy than we create. Furthermore, not only is it possible the energy form collected via fusion may be useless, it may also be dangerous. Given the current state of the art, storing it could be equivalent to storing the universal solvent. It's one thing to place radioactive wastes in lead containers and bury them; it's quite another to bury extremely strong but unusable potential.

Once we recognize how both fission and fusion function primarily as filtration processes, we can see how aligning the pores greatly decreases the amount of energy needed to initiate the process. Think of the energy necessary to overcome inertia as that necessary to align an unruly group of individuals before commencing an orderly march. The more unruly they are, the longer it takes. Obviously the *more* coherent the group is at the beginning, the less energy it takes to get it moving. Compare lining up fifty raw recruits versus the same number of career military personnel. The leader of the former must work a lot harder to get his or her group organized and ready to go.

Heat and Resistance

From our discussion this far we can see how the philosophy of initiating energy is really the crux of this principle. And although we've talked about how fission and fusion work, we still have a disturbing "why?" to deal with: *Why* does it take so much energy to initiate these reactions? We already know both fission and fusion use a great deal of energy to overcome inertia. This must mean that systems are *resisting* the reaction or activity; and resistance to anything means there's friction. Although any reaction between any forms—be they 7α, molecules, minerals, people, or nations—may be forced, the entire system by virtue of its resistance is saying something isn't right, something needs to be changed. So perhaps the real problem isn't finding something other than heat to initiate fission and fusion reactions, but rather to determine if we want to use such reluctant unnatural reactions at all.

However because the belief that high heats of reaction are necessary is so deeply rooted in technologocial philosophy, this belief must be unraveled very carefully. Some of our current philosophy stems from more ancient beliefs; we often speak of the glory that comes from hard work, the sweat of one's brow and the heat of battle. Think of all the nonproductive effort that goes on in most large organizations—meetings, committees, reports, studies, memos—just to get things going, to produce minor changes or none at all. So often we believe that only through massive amounts of energy expenditure can anything be accomplished.

Less is Best

So in order to disentangle ourselves from this highly structured and rigid belief system, we must recognize that hard work may not only make Jack dull, but also inefficient. However it's not the work that makes Jack dull; it's his belief that his work must be hard. We get out what we put in. The streaming effect is based on the premise that like begets like. If we wish energy out of a system, we must put some energy in. If we put the right energy in, the reaction begins and then continues on its own.

Our belief that we need large amounts of energy to initiate energy-producing reactions is well mirrored in many parallel situations in Western society. "All beginnings are difficult" and "More is better", we say. If we put aside $10,000 for a rainy day, it may not be long before we believe we need $20,000. If we attempt to move an object with some force and the object fails to move, if our subordinates don't respond to our ideas, we apply more force or speak louder. In warfare, a successful attack must not only destroy the enemy's industrial capability (good); complete obliteration of cities and their civilian population is necessary to prove one's power (better).

What's interesting is that any civilization can only maintain such an orientation so long because the bigger-is-better isn't only impractical, unfulfilling, destructive and eventually boring, it's also self-limiting. It's a variation of the well-known behavioral principle called flooding and is equally applicable in physics. The more we stimulate a system, the more numb it becomes to that stimulus. Think of how you jump the first time your nephew bounces his ball against the house; however as the noise continues, you're able to ignore it. He can try to keep your attention by making more and more noise, but it's a lot easier for you to ignore him than for him to come up with the necessary energy for sustained hard bouncing. Once you ignore the behavior it stops because you no longer acknowledge its existence, whether it's there or not. If your nephew is bouncing the ball to irritate you, once you ignore it he has no reason to continue and will probably quit on his own after a few more bounces. In such a way the boy's belief that if a small bounce was irritating, a louder one would be more so becomes self–limiting.

Similarly, the first car was designed to replace the family horse, then ten horses, then up to five hundred horses and more. At that point people began asking themselves, "What in the world do we need all this power for just to go to the supermarket?" Suddenly the big cars plunged into disfavor becoming "road hogs" and "gas guzzlers" and their owners unpatriotic energy wasters. Given these and other examples from your own experience, it shouldn't surprise any of us if the current flood or more-is-better beliefs inundating computer and other electronic technology as well as our defense system goes the way of too large cars and irritating bouncing balls.

To be sure, there will always be that minority concerned with such things, but the more-is-better premise is highly unproductive and has little to sustain it.

Similarly, the belief in the need for high initiating energies for fission and fusion reactions is undergoing an equivalent to the psychological process of flooding and the neurological phenomenon of fatigue. Once the system or belief is driven to the limit of "more", it becomes virtually paralyzed. Within the behavioral context, individuals who function under flooding first experience great peaks (elation) and then valleys (depression); as long as there is more to add they feel good. When they reach their limit of more, they plunge into depression. Your nephew chortles gleefully the first time you jump at the sound of this ball; then he delights in seeing how *long* he can maintain your attention. When you fail to respond, he becomes angry, perhaps even pounding the ball or throwing it as hard as he can. Finally he quits and sulks, says he's bored and wants to go home: "It's no fun here. There's nothin' to do."

Similarly, nerve endings subjected to increasing amounts of the *same* stimulus soon lose their ability to respond at all. We've all experienced olfactory fatigue—we enter a room or building and become aware of a certain pleasant or unpleasant odor which quickly fades. Another person entering a few minutes later reminds us of the odor by commenting, "Wow, what stinks in here?" And although we realize we must have light to see anything, we've all experienced blindness caused by too much light. The amount and duration of any damage in these cases is directly related to the amount and duration of the excess.

In a fission reaction the same process is taking place, only it's not permitted to stop. The system is being forced to respond to an initiating flood of energy far in excess of its needs. Not only must the lattice structure struggle to maintain *itself*, it's also struggling to handle the excessive load of energy in the form of heat being forced through the system.

When a nuclear reaction must absorb an initiating energy surge and is then forced to continue in the resultant unnatural environment, the uranium lattice structure becomes "numb"—so distorted and structurally altered it passes virtually all forms of $_7\alpha$ through it and loses many in the process. In essence, the system gets out what was put in—incoherent heat—only it gets out less. Compare this to preparing a basic boiled dinner with meat, vegetables and seasonings. At the end of the process you want to strain your stock to separate the meat and vegetables from the liquid so you can serve the former and make a tasty sauce from the latter. However, suppose your family and guests are clambering for food immediately. Rather than letting the mixture drain through a sieve at its own rate, you dump it all in at once and stir it vigorously with a spoon to hasten the process even more. You press so hard the sieve openings become distorted; some even break. At the end of the process you have less than what you started with; you have no recognizable meat and vegetables, nor clear stock with which to make a sauce. All you have is an indistinct mass of something—part of it clinging to the sieve, the rest in a bowl—which may include tiny pieces of broken wire from the sieve.

Understanding why both amount and purity or coherency are necessary to prevent flooding and the breakdown of the lattice mechanism is one thing, but how can we determine the least amount of energy necessary to initiate a reaction? Obviously we can use trial and error "less is best", reversing the more-is-better orientation. Another way is to take the square root of the smallest amount of *perceivable* lattice matter. Because we know it takes 10^{13} $_7\alpha$ pair per cubic millimeter (mm^3) for perception,* we can initiate 90% of all energy-filtering mechanisms with

$$\sqrt{10^{13}} \;_7\alpha \text{ pairs or } 3.16 \times 10^6 \;_7\alpha \text{ pairs.}$$

Having established the *amount* of energy necessary to initiate a fission or fusion reaction, how do we establish the *form*? From our discussion we know the best way to prime any system is with the form we want to collect; we don't prime water pumps with oil

---

*See *A Primer of Rotational Physics*

or maple syrup, for example. Therefore, priming with an incoherent source sets the lattice to collect incoherent forms because the primer or initiating energy sets the stage. We get out what we put in—garbage in, garbage out: incoherency in, incoherency out.

Does this mean we must put pure coherent heat, pressure, light, sound, or potential in to get it out? Not necessarily: but certainly the *more* coherent the initiating energy form, the *less* required to prime the system and the more efficient it becomes. Wouldn't the creation of such refined energy sources require the expenditure of equal or greater amounts of energy? Again, not necessarily; in fact, it may be accomplished so simply as to sound nearly ludicrous. Although beyond the scope of our discussion in this book, it is possible to pass an incoherent light beam through a naturally changing water lattice thereby creating nearly-coherent heat and pressure. The initiating source of light is placed at right angles to the *passive* (unforced) motion of a piston which causes the water lattice to expand and contract. In such a way the incoherent light is initially filtered by the water, providing a much purer form of natural energy.

We can think of incoherent light as similar to complex proteins and that light's component energy pairs as amino acids. It's much simpler to break down proteins when they're dissolved in a solution. Similarly, by passing our light through water we can more easily break it down into much smaller blocks of $_7\alpha$ which can then be filtered and recombined into more coherent forms. This is a fission-fusion reaction in its simplest, safest, cleanest, most efficient and natural form. In the case of light and water, the mechanism is such that the initiating energy is barely distinguishable from the maintenance energy required to keep the system running. Although this may also be said of current fission-fusion mechanics, this occurs in the light and water system because the levels necessary to initiate and maintain the reaction are equally *low*, not high.

The low levels required to initiate and maintain the system are philosophically much, much more difficult to accept than extreme high temperatures and potential explosions. As a group, we civilized folk are very dependent on external manufactured energy to supply excitement in our lives. Cut off from electricity for lighting and appliances or fuel for heating and combustion engines, most of us find life dull and boring; some even find the concept frightening. Not only is the idea of being without man-made energy frightening, the idea of being without light or heat in the middle of a house-shaking thunder and lightning storm can be downright terrifying. Why? Because as dependent as we are on manufactured energy, many of us still feel uneasy in the presence of *natural* energy. Surprisingly many of us feel we have more control over a light switch or a vehicle than we do over ourselves and our lives, which includes all natural energy forms. Although we intuitively know our natural power far exceeds any inherent in any manufactured system, we give our cars, TV sets, computers power to cure our boredom. However rarely do we ever relinquish the awareness that, if such systems break down, we're still ourselves. We can choose to walk away and create something else, whereas our machines must sit there having no function without our presence regardless how much manufactured energy flows into them.

Summary

Summarizing this principle, we can say there are two major beliefs springing directly from humankind's awareness or attitude of the self that affect our energy philosophy. One is that the greater the initiating force required, the greater the output; the second, that the initiating force always exceeds the maintenance force. Because of these beliefs, systems producing a great deal are invariably seen as having to overcome a great deal to get started. We need only look at the number of great leaders blessed with charisma, intelligence and intuition who felt the need to suffer, creating physical ailments from chronic back pain to paralysis, constantly courting personal disaster via extra-marital liaisons, experiencing the tragic death of loved ones, and so on. Because their natural ability was so great, they couldn't tolerate how *easy* it was for them to do things. Therefore they deliberately messed things up, made *conscious* misjudgements to prove they were as fallible as anyone else; and in so doing they maintained the myth that energy/love creation is difficult and requires great sacrifice. We need only look at the lifespans of such individuals and the manner of their deaths to realize they fought their intuitive awarenesses right to the end. When they acted as complete fools, tempted disasters to denigrate themselves and their powers, it still didn't change their power. It's like trying to run from one's own shadow, instead of realizing one is the *source* of both the light and the shadow. To be sure, we can eliminate our shadows by getting rid of the light; but then all is darkness. However, we can also eliminate our shadows by moving closer to the light; and at some point as we move closer we invariably must discover we *are* the light.

The reason it takes so little energy to initiate a pure energy filtration system is because even the very first $_7\alpha$ begins the primary process. It's the first photon through the black hole; the first visitor in a strange land. Once the first of the various $_7\alpha$ forms pass through appropriate lattice openings, others of its kind find it much easier to follow. It's not necessary that the leader create or distort the lattice, fight its way through, or sacrifice itself. All it must do is pass through and the rest will follow. The more bulky, less pure the leader, the more likely it is to change the lattice. The more the lattice changes, the less likely it is that anything it filters will be the same as that which came before it. The leader who does the best job is the one whose followers are so numerous and similar we can't distinguish the initiator at all. The most efficient reactions are those using minimal start-up energy and producing no waste in any form.

7 | The Fourth Principle

The Fourth Principle: Any chain reaction is unnatural.

To understand this principle, let's begin by defining a chain reaction as one which is a function not only of numbers but also of time. With each successive generation the amount of time necessary to change decreases, usually exponentially, whereas the number of entities reacting increases, also usually exponentially. Think of the word-of-mouth or pyramid effect: One person tells two others who in turn tell four others, etc. If it takes each group half the time to relay the message as the preceding one, we have a true chain reaction. Compare this to a single line of dominoes where the time interval between contacts is constant and the contact is always one-to-one.

Nuclear fission is a chain reaction; each successive change involves a larger population and requires less time per individual change. Does this form of energy seem natural? Think of having your normal intake of food gradually increased in both amount and number of meals a day. Do the same with gasoline for your car and electricity for your appliances. Not only is energy in that form unnatural, it's also destructive. The simple fact that our toasters and washing machines can't respond to the timing and magnitude of chain reaction energy serves well to demonstrate man's innate awareness of the fallacy of such systems.

Chain Reactions and Energy Production

To see how chain reactions effect energy production, let's compare our lattice work to a series of toll booths on a crowded turnpike.

There are two ways we can increase the number of vehicles going through the system:

- Use one booth and increase the speed (decrease the time interval) of the vehicles passing through the booth and the rate of the personnel collecting the tolls, or
- Open up more gates and maintain the original rate of vehicles and personnel.

Current fission-fusion mechanics is based on the former approach. To be sure, by increasing the velocity of the cars through the gate as well as the rate at which the attendant collects tolls, more cars can be processed. However, we need only work the scenario through at ever-increasing rates to realize

- The performance of the total system is determined by the *slowest* participant.
- One failure or breakdown of either vehicle or personnel ties up the entire system.

In theory, we should be able to predict the behavior of our vehicles as well as their velocities; the truth is we can't. As we increase the vehicular velocities geometrically, the drivers are less aware of change and have less time to respond to it even when they are. Similarly those collecting tolls become so harried they can't keep track of what's going on—they overcharge, undercharge, drop money. As the pace increases, both drivers and gatekeepers become more and more strained and invariably the system breaks down; a toll collector loses his cool or there's an accident—anything to put a stop to the unnatural pace. Given such characteristics, is it any wonder many view nuclear reactors as rather fragile devices?

Because our turnpike only has a single toll gate, what occurs beyond the gate? At most major turnpike exits there are usually a number of choices we must immediately confront: Exit 5N or S? Interstate 94? South into town? Toward the coast? If there's a line of rapidly moving vehicles behind us, there's little time to make these choices; our main thought is to get away from the mess.

Furthermore, if our velocity is sufficiently great we can't even pull over. Even if we could, how could we ever get back into that rapidly moving line of cars full of frantic drivers? Thus as the drivers come through the system their main goal is to get out of it. All former plans of ultimate destinations are abandoned; those initially heading to the beach find themselves careening toward town because it seems an easier or faster way out of the mess. Regardless of the route taken, all drivers find their velocities excessive and must fight to decelerate and maintain control as they exit. Even if the system functioned perfectly and the trip was accomplished in half the time or at four times usual speed, we need only imagine the exhausted, trembling driver sitting in his overheated machine in a town where he doesn't want to be to question what is gained. Of course if the system breaks down before this, the results are equally questionable—the difference being that between the high morbidity of the end-product versus the high mortality of a flaw within the system. The driver who arrives is alive and in one piece, but the experience has been sufficiently upsetting to ruin his entire vacation; those who are involved in accidents occurring in the high speed process may never reach their destination at all.

Now let's open more gates and maintain the velocity at an even pace. The same number of cars pass through per unit time as in the other system, but if one car or gate has trouble, the others continue functioning. Because the velocity isn't excessive, drivers have time to choose the proper roads as they emerge from the toll gates. So if we want more energy to flow, we don't increase the system's velocity unnaturally—we open it up more and let it relax.

Recognizing Natural and Chain Reactions in Ourselves

Using this definition we can see that the unnatural status of chain reactions is the result of their timing more than their content. *Anything,* regardless how alien or detrimental to a system, may be absorbed with minimal or no permanent damage if increases are accomplished via increased flow or "openings" rather than increased flow per unit time through the same channel. The individual who totally responds to the adrenalin surge associated with fear opens all channels for *maximum* action whether this manifests in flight, fight, or freeze defense responses. Pupils dilate, heart rate increases, involuntary noncritical functions such as digestive processes slow, the mouth gets dry, all senses (vision, hearing, taste, smell and touch) become more responsive to stimulation. The individual who attempts to abort the natural adrenalin response by restricting it to only certain areas winds up sending dangerously high levels of energy to a relatively narrow receptor system. Furthermore, because the natural response to stress is to dissipate it throughout all systems, channeling that energy to a single system such as the involuntary one controlling the muscles lining the circulatory vessels or digestive track wreaks havoc with one's health. In general the adrenalin sends the message "Get ready to do something!" When the body readies itself and nothing happens or when the energy is channeled disproportionately into a few systems, the body becomes trapped in a severe physiological and emotional contradiction. Rather than *confront* the fear by forcing it on one involuntary physiological system, it's far better to dissipate the fear entirely. So the person who becomes frightened can respond by creating emotional, intellectual and physical outlets to dissipate the fear. Compare responding to a frightening situation by permitting yourself to logically analyze it, unashamedly emote *and* freeze, fight or run versus allowing yourself only one of these outlets and bottling up the rest. Like our turnpike with only one toll booth, the system soon breaks down from the unnatural strain and resultant chain reaction.

One of the most common chained sequences occurs with anger which builds in intensity. The almost palatable tension so created tells us immediately it comes through a most narrow channel indeed. This occurs because not displaying irritation is part of our "civilized" fabric—a sign of self-control. Such tension-creating responses can work as long as that undisplayed anger is forgotten. However, it rarely is; it's stored and added to, like our ever-faster single line of vehicles getting closer and closer to that one gate. Like our cars, eventually the rate and velocities of the minor irritations carry them into the realm of a full-blown assault and the human system gives way to uncontrolled rage.

Do you think you can create a chain reaction? Try this thought experiment. Imagine your most troublesome problem and work it through by seeing the solution as a series of linear events occurring one after another. For instance, suppose you need an operation to correct an annoying but not life-threatening medical condition. You don't have health insurance nor enough money to pay hospital and doctors' bills. One solution expressed linearly might look like this:

1. Get a job whose benefits include employer-paid health insurance.
2. Stay on the job long enough to be eligible for benefits and build up vacation time.
3. Visit your physician.
4. Set up a time for the surgery.
5. Request the vacation time.
6. Prepare for the operation.
7. Be admitted to the hospital.
8. Have the operation.
9. Recuperate.
10. Begin to live life in your improved condition.

In this case we can represent this sequence of events like this:

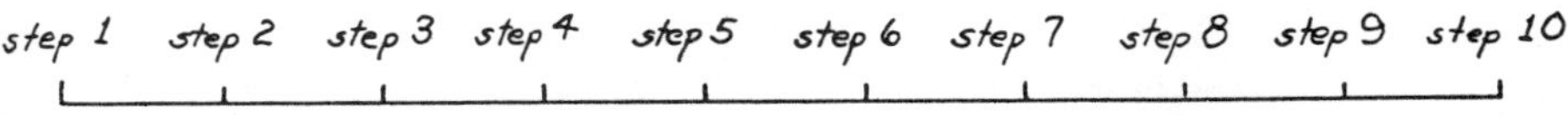

To convert this to a chain reaction, we simply import more and more velocity to successive steps which simultaneously shortens the time between them:

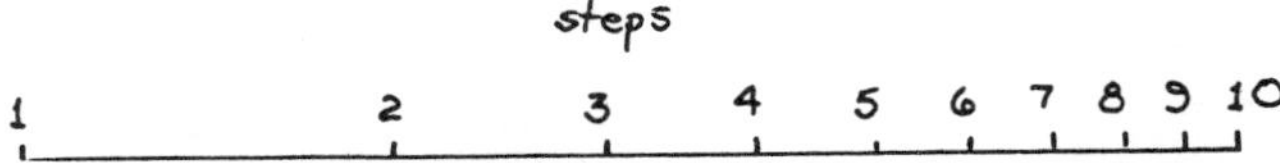

Now let's take the same problem and its component parts and

deal with each one separately giving none greater speed, credence, or power than any other. Instead of seeing the steps linearly as a function of time, they now appear simultaneous:

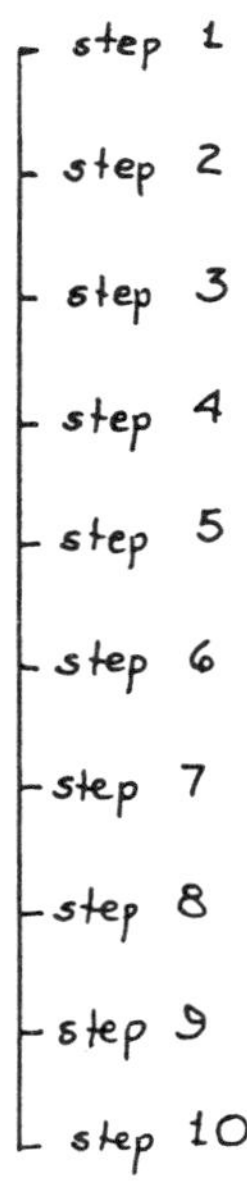

Instead of viewing getting a job with paid health insurance as part of the process of getting from Step 1 to Step 2, we now see it as a meaningful change in and of itself. If you take a job because you want to be able to afford knee surgery rather than because you want the job, how good are your chances of succeeding in that job or lasting long enough to earn the necessary cash or vacation time for your surgery? Similarly if you haphazardly prepare for the surgery or rush recuperation, you decrease the likelihood of results that meet your expectations.

Although weighing each step equally sounds simple, many of us find this difficult, especially when dealing with problems and solutions we consider important, those having strong positive or negative connotations. If we anticipate some favorable result, we want to rush through the preparatory stages; if the impending result is perceived as negative, we want it over as soon as possible. In both cases by rushing through the process at an ever increasing rate rather than appreciating each step as a complete and valid entity, we deny ourselves a great deal of knowledge and experience.

Isn't viewing each step as equal the same as getting bogged down in process? Not if you consider each step as contributing its own result as well as being part of a greater process. Compare the student who rushes through college, medical school, internship and residency because he or she is so anxious to be a surgeon and enjoy a surgeon's income versus those who take advantage of the special opportunities and contributions each phase has to offer. Which person gains the most from the process? Which one makes the most efficient use of his or her time and energy? Which one is more likely to respond to the needs of, and communicate with, the larger population? Because we believe the principle of chain reaction dominates much long-established thought regarding energy production, we urge you to work through these experiments so you can take any problem and evaluate the solutions as a series of linear, chain, or simultaneous events.

Openness and Flow

View your own beliefs and their emotional, mental and physiological results as infinite cars approaching infinite gates. If we're rigid and inflexible in our beliefs, only a few of those gates are open. When we encounter others and want to transfer one of our beliefs to one of their gates, how probable is the alignment between the two of us? The more open *we* are, the more likely a successful exchange will occur.

How would you estimate the long-term success of political, scientific, or religious figures who demand all their followers think and believe exactly as they do? Contrast this to the person who serves more as a guide, helping others develop beliefs that are right for each unique individual.

If we're open, even if others are rigid we can send out so many vehicles at an unthreatening, regular pace, the chance of encountering even their singular open gate is much greater. Furthermore, if our gates are open we're much more likely to be able to receive what they offer us. Compare this to two individuals of fixed but different beliefs encountering each other, each firing a stream of ever-faster cars into the other's closed gates. Is it any wonder that anger and destruction is often the only result?

Does this mean all chain reactions are grounded in evil or negative emotions and thought processes? Not in the least. Both forms of reactions or responses are merely *processes* and as such may embody any or no emotion as we see fit. To see how this works, let's apply the chain and natural reaction processes to two common emotions: hate and love.

First let's create an adversary of singularly unpleasant character, someone who has thwarted your every attempt at personal fulfillment, has what you want and taunts you with it. Let's further give this individual personal characteristics and habits which you find totally disgusting. Finally, let's place this person in a position that frustrates you—a clerk who makes you wait in line, a supervisor who has you perform meaningless tasks or demeans your best abilities.

Now that we have a villain, let's apply our two forms of reactions. To initiate a chain reaction, take one event or characteristic (his beady eyes, her nasty manner) and begin building on it. As you add each negative judgment, do it at twice the speed or in half the time as the one before. What happens? Pretty soon the negative "cars" are moving so fast you have no time to define them and you merely add raw emotion or incoherent energy to the system. This common phenomenon leads to arguments and feuds which go on for years and years with no one remembering the original cause. Bear in mind that when thinking is uni-level or "one gate", not only are the chances of that thinking being transmitted to the corresponding gate in another quite slim, it will also be perceived as threatening because of its ever increasing rate. Think of thoughts as a batallion of soldiers leaving a gate in one fort and charging the gate in another. Regardless whether the motives are good or bad, a batallion charging ever faster and faster needs some place to go. If it encounters a gate in a straight line with the one just exited all is well providing there's a place for all the soldiers to go. If there's no aligned gate, the soldiers slam into the wall, lose their alignment and dissolve into an incoherent mass.

In addition to recognizing chain reactions by their inherent increased velocity and tension, we know we can also recognize them by their process orientation. Thus in our emotional hate chain reaction, the individuals involved rarely have any idea what result they want. If the process of hating blacks, Jews, nuclear energy, war, abortion, taxes or pollution is all there is there's no goal, nothing to do but hate more.

If we look at the results of chain reactions, whether in the arts, sciences, or personal relationships, we see that what keeps them alive can *only* be the process, the thrill of the hunt or chase. This occurs because the decreasing time span and increasing velocity doesn't allow us to appreciate each step, its advantages and/or consequences. Because of this, chain reactions invariably require more energy and produce less than natural reactions; furthermore, they invariably produce waste.

Having created a hate chain reaction, let's do the same with love and observe the physical, mental, and emotional results. Like anger or hatred, love manifested in a chain reaction loses meaning as the velocity increases. As velocity increases, tension increases and the major motivation becomes a need to get rid of the tension. Although some may equate this with showing or making love, it's more akin to the feeling one has when one needs to deficate or urinate—a desire to relieve an uncomfortable feeling of being "too full". A point is invariably reached in any chain reaction where the identity of any inciting factor is lost in the momentum of the reaction, as well as that of any perceived goal or result.

Now let's run our arch villain and perfect lover through a natural reaction. Bearing in mind we want each aspect of that individual to carry the same charge or velocity and fill all our gates, we may practice and experience this phenomenon in several ways.

First we may take all those things that make us hate or love a person, assign them *equal* value and mentally take all these simultaneously through gateways at the same time. What happens? Is it possible to concentrate on the individual characteristics that make this person so hateful or lovable? Can you keep track of beady or beautiful eyes when numerous other equally-weighted traits are simultaneously moving?

Having dismembered the villain and lover into equal packets which we feel justify these emotions, let's expand the number of gates. This means that instead of having ten *reasons* to hate or love this person we now have a hundred, then a thousand, and so on. What happens? Just as we eventually reach a point of giving up trying to distinguish the reasons in the chain reaction because the pace becomes so frantic, so we give up trying to distinguish the "why" in the natural reaction. However, the result and its effect are quite different. With each opening bank of gates, more energy flows; simultaneously, rather than experiencing a feeling of increased tension we feel more and more relaxed and relieved. Even if the object of our hate or love isn't receptive to our energy, it makes no difference. For example, suppose you say you hate blacks, Jews, Catholics, homosexuals or any other group you consider sufficiently different from you to be threatening. If you only have one gate—"I hate him because he's black."—*everything* that person does must be channeled through that gate. Even if he does something that would please you if someone of another race did it, you must perceive it as a hateful act. However, if you create more hate gates— "I hate the kind of ties he wears.", "I hate the way he drives.", "I hate the brand of cigarettes he smokes."—you create a completely different system. Although more gates would seem to imply the opportunity to hate more, because all those gates function simultaneously, the overall impact in each area is lessened.

Even though this would seem a beneficial way to diffuse *negative* emotions or reactions, doesn't it also serve to dilute positive ones? Not in the least. The problems that arise in our physical, mental and emotional well-being aren't the result of the messages *others* choose to send to or receive from us. Our problems arise because of what we choose to send out and receive *ourselves*. If we withhold or send only through a limited number of gates, we're bound to create internal tension. If we try to channel all we receive from others through equally limited or defined gates, we experience tension we attribute to outside forces. The person who expects another to behave in a specific way "if you love me" is no more open, receptive and giving, than one governed by blanket hatred. In both cases the individuals maintain their orientations based on their interpretations of what they believe the other is saying or doing, never realizing these interpretations result from their own openness or lack of it.

Another way to look at problems is to envision them as having a cellular structure just like any other living thing:

On this level we can once again see the magnitude of stress created in a chain reaction compared to a simultaneous natural one. Even if one cell moves at a scant 10^{-10} mm/sec, if that velocity doubles with each successive cell division, the velocity and tension can mount quickly. Considering there are millions of cells in the body, the total effect can be phenomenal. However using the principle of natural reactions, *all* cells move simultaneously; instead of each cell moving at an ever increasing rate relative to its predecessor, all move the same amount in the same time. The total system moves but each part remains stable relative to each other.

By opening up a system fully, do we create or do away with energy or emotions? If we view emotions as a form of energy we can see that how we feel about *anything* can only be a projection of how we feel about ourselves. We find loveable or distasteful about others what we find loveable or distasteful about ourselves. If the idea of becoming more and more open and relaxed until your limiting definitions of anger, hatred *and* love vanish strikes you as singularly dull, try it. You won't become an unemotional, unfeeling, uncaring blob—far from it. By learning the difference between manufactured chain-reaction energy or emotion and the real thing and opening yourself up so that natural coherent forms flow more easily *in both directions,* you increase your range of experience tremendously.

What many of us equate with our spiritual and emotional state are primarily the incoherent sources and products of conscious chain reactions. What we begin and end with in a natural reaction is far, far different and well worth any effort needed to develop the technique. Practice is the only way, always testing yourself with the questions: What is my product? What did I get for all my emotion, thought, and work? If the answer is "Nothing worthwhile or lasting," chances are you were involved in a chain reaction, process orientation, yielding more awareness of process, little of result.

By becoming consciously aware of natural processes versus chain reactions, we can develop a skill that enhances our dealings with others immensely. Furthermore, the awareness of natural reactions and their advantages in our personal experience can then be used to enhance our understanding of how mechanical or energy-creating devices adhere to or violate this principle.

Now that we recognize how chain reactions differ from simultaneous natural ones and how we can create either form within ourselves, let's take a closer look at energy.

8 | The Fifth Principle

The Fifth Principle: Energy in one form can supply all the needs of the earth.

This principle refers to *pure natural* energy with its human emotional counterpart, *pure* love. Note the use of the word "can" in this principle; we can use one form of energy to supply all needs, but we don't have to. We can employ any of the pure energy forms—potential, light, heat, sound, or pressure—or combinations of these forms. Furthermore, we could add the word "forever" because this pure form of energy uses no resources (coal, oil, gas), creates no unwanted byproducts (unusable heat, smoke, chemical compounds), and produces no waste.

If we put the idea of a pure energy form in the context of the simultaneous flow of the smallest $_7\alpha$ units described in the last chapter, we can see how this must be. If there is a smallest entity and it retains the ability and freedom to change, then anything composed of that entity must have that same ability. Furthermore, any single entity within a group or greater rotation must be interchangeable with any other.

Imagine we built a working computer out of Tinkertoys® as one computer wizard did. Only instead of using different sizes and shapes, we create all the computer's parts from the same basic unit. We can take a unit from the keyboard and interchange it with one from the disk, chips or power source. These units represent the pure form, the source of our computer; regardless what part breaks down, we can repair or replace it using the basic unit.

If we want to alter this structure, does it make sense to hit it with a crowbar or detonate 100 kilograms of dynamite in hopes of changing the desired units to save time? True, we might save time by decreasing the time necessary to take the structure apart, but we would also destroy many of the units in our haste. Wouldn't it be better to dismantle the system, basic piece by basic piece, even though it might take longer? By doing so we could group or rearrange the pieces any way we want.

If the time saving, albeit destructive, approaches sound familiar that's because the two premises dominating technological thought today are time-oriented. We want to produce energy, information or cars as quickly as possible, adding the principle of "faster is better" to our previously discussed "bigger is better". In doing so the "more of the same" philosophy expands becoming "more of the same even faster".

Suppose an engineering student looks at our computer and thinks "Golly, if I could knock that thing apart I could build a calculator to do my math homework faster." Then he destroys the computer and builds his machine. However just as he finishes, his teacher comes along and sees the remains of the computer and the student's creation. Rather than praising his student's ingenuity, the professor is furious. "You fool! Didn't you know that computer could do your simple math in addition to countless other tasks? Now look what you've done! Because you were so ignorant of the potential of what you had, you destroyed it to create something much less."

In our story the student's objective is to create something that will expand his ability to do his math homework. He's perfectly capable of doing his math on his own, but he believes a calculator will enable him to do more homework faster. Compare this to our creation of electrical or fuel-powered machines to cook, sew, wash, dig and travel faster than we can ourselves. Any refinements to these systems usually involve the use of more electricity or fuel to do the job "bigger and faster." Rarely in such thinking is the idea of doing the task a completely different way considered. (Our student never considers developing other mental skills to replace counting on his fingers, for example.) The same thing happens when energy production is the goal of the process or machinery. We take the basic idea that energy is "released" when certain substances are compressed or shattered and merely create the refinements that permit more release faster. Because we only acknowledge the released form, we can't recognize any other forms which we destroy in process, much as the student destroys the computer to build his calculator. Furthermore, because we're well aware of the limitations of our acknowledged form, we can't believe it can ever fulfill all our needs because we know its creation depletes our natural resources. Therefore in addition to being trapped in a philosophical and technological chain reaction, we're also caught in a vicious cycle.

Changing Our View of Time

How can we train ourselves to view energy forms and their creation in a different light so we can consider other options? Because all chain reactions and therefore all current energy production to one degree or another are time dependent either via the production itself or storage, let's talk a bit about time. In the last chapter we created a thought experiment in which we simultaneously passed an infinite number of entities through an infinite number of gates. We became aware of the greatly decreased resistance inherent in such a system, but also of its dilution of power. For some reason, being to the side of an infinite row of troops marching perfectly aligned seems less powerful than leading a charging force at ever-increasing velocity, even with the latter's likelihood of total breakdown at any instant.

What's the difference? The answer has to do with time. If we believe time is an absolute in this reality rather than that each creates his or her (or its) own time, then time becomes our god, our measure of change. The more time that elapses, the more we change. The interesting paradox is that it's not the time that creates the change; it's the change that creates the time. That's why we invariably seek to change something when we're bored and time seems to pass too slowly. If change were a *result* of time then all people would tend to respond to the same situation the same way. At the end of a half hour all the people observing an event would have experienced exactly the same thing. We need only place a three-year-old, an avid sports fan, a sports hater and a teenager listening to rock music through earphones in front of a television broadcasting the Super Bowl to appreciate the individual differences. The three-year-old's half hour has no meaning adults can understand; the fan experiences an interval more like a few minutes; the sports hater feels the game's been on for hours; the teenager gauges time from his music.

Although we know likes create a streaming effect, we also know the effects of boredom. It's as difficult to be around a bored individual as it is to be bored ourselves because that person is extremely time-sensitive without being time-stable. Bored people are well aware of the *passage* of time, but they don't like the interval. They attempt to change by literally speeding their seconds and minutes, hoping to find that rate most likely to create the desired result. Similarly, those who are nervous or high strung often feel there aren't enough hours in the day for all their desired changes to occur. They respond by attempting to slow time down. In both instances the only changes are those physiological changes we would expect when our biological clocks are altered—primarily variation in heart-rate, blood pressure, respiration and digestive transit time.

Time, Tension and Resistance

If tension and resistance are functions of time in human reactions, do these same relationships exist in energy-creating reactions? If our principle is truly rotational and therefore unifying, they must. To understand this, let's create a natural energy system with $_7\alpha$ simultaneously exiting gates at uniform velocity. We already know there's no tension or resistance at the gates by virtue of their infinite number and no time since all exit simultaneously. However, what occurs between the time the $_7\alpha$ leave the gates and arrive at their destination?

Again imagine a lateral line of soldiers and a single target in the distance. First, let's look at the probability of hitting that target: We have a much greater probability of hitting the target using a natural system. In fact, in the natural reaction the probability approaches one whereas in a chain reaction it approaches zero.

Compare the following:

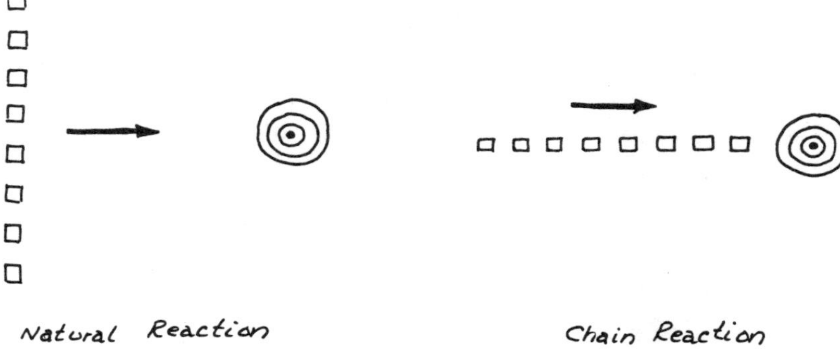

Regardless where the target is, in a natural alignment the chances of at least one $_7\alpha$ finding it are excellent. And, if one finds it, the steaming effect takes over and channels the others:

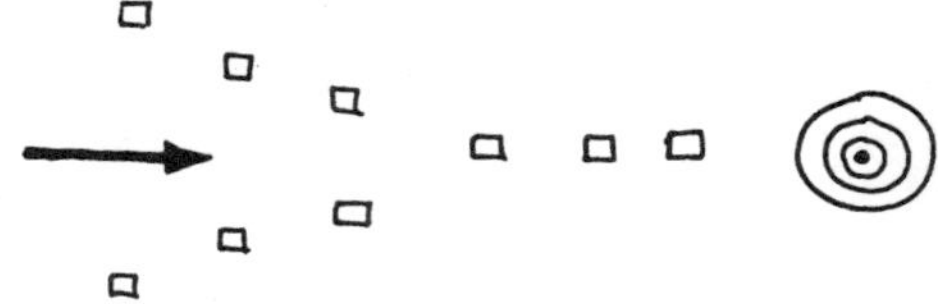

Unless the lead $_7\alpha$ is aligned with the target in the chain reaction, the chance of any $_7\alpha$ hitting the target is nil.

It is inherent in the natural flow of $_7\alpha$ to realign in the most favorable configuration following the path of least resistance. This can *only* occur if they aren't experiencing tension or resistance to begin with. Think of a group of soldiers spreading out to find the easiest way to approach the target. As soon as one finds it, the others quickly abandon their own paths in favor of this one.

Tension-resistance is a product of time. If there's no time, there's no resistance. Because all the $_7\alpha$ are simultaneously present, they can immediately determine which corresponding gate is open. As soon as one $_7\alpha$ enters, the rest respond to the void or relative vacuum produced by the "loss" of one $_7\alpha$ in their ranks and follow. Such can only occur when all are aligned and moving in the same direction at the same rate. To be sure, if the leader of our ever-faster chain reaction enters the gate, the entire column will also follow; however, we know the probability of this is relatively slim to begin with. Furthermore, even if it does occur we must still contend with the high probability some or all will be hindered by tension- or resistance-related events caused by the ever-increasing rate.

Simultaneous States

Now let's consider this principle in terms of the inherent $_7\alpha$ pairs we know comprise pure energy regardless of its form. As long as what we collect and utilize is inherently aligned, the source is inexhaustible because it's never used or permanently converted like wood, oil, or U-235. However, we can never separate the source from the consumer. Obviously in rotational theory nothing can be *only* a source to something's or someone's sole consumer status. All must function as both source and consumer, giver and receiver simultaneously. We can never take without giving back or give without taking, whether we choose to consciously acknowledge both aspects or not. Thus a fission reaction, relatively speaking, begins with incoherent energy and coherent mass and produces coherent energy and incoherent mass. Rather than view such dichotomies as vast differences, pendular extremes or heavens and hells, see them as opposite sides of a record being simultaneously played by two needles, each needle located in the same groove.

Thus these opposites progress simultaneously at their own pace. If "good" occurs in your life, you have the added joy of knowing the corresponding ante-matter (note, not *anti*-matter) "bad" state has simultaneously been experienced; alternate probabilities, regardless of their translational time, occur in the same span afforded the conscious or perceived act here. For example, if you receive a fine piece of news in the mail or hear an excellent concert, all other probabilities or states of those events are *simultaneously* being played out. Even if within the unrealized probabilities there's one where the letter's contents or the concert is wretched, it's realized in the same time it takes you to read your pleasant letter or listen to your pleasing concert.

Therefore, those who constantly expect the worst aren't that far off in their intuitive understanding of the actual process for if the worst occurs, the presence of the best is guaranteed. However if we view the two events linearly instead of simultaneously, we split the matter and ante-matter or alternate probability states and expect one to follow or precede the other. Consider this mathematically: When we evaluate a probability it's firmly grounded in time, either a single moment or a time span in which all probabilities within the system occur. As soon as one alots a different time to a probability, it's no longer part of the original set and gives rise to its own. For example, Ernie Hower gives himself five years to become Gardenia Electronics' general manager. Included in this scenario are other probabilities which could occur in that five year interval:

- He remains a junior manager.
- He's promoted to another area.
- He's fired.
- He quits.
- Gardinia Electronics folds.
- He dies.

If by the end of the five years Ernie is general manager, all the other probabilities vanish, or rather should vanish. However if Ernie continues thinking about them, they take on their own identity and time frame.

Psychologically and in terms of spiritual energy, an understanding of the *simultaneous* manifestation of all probabilities is most freeing. If we recognize our conscious *choice* to experience the absolute worst of something, we *know* the experience is being shored up on other levels in a much milder and even quite pleasing form. Similarly, if we experience what we consciously acknowledge as a good or positive event, we also know there's no "worst" backlash or component which *must* be dealt with at some later time.

In other words, there is indeed punishment for every crime, but it occurs simultaneously with the act itself. We indeed reap our rewards or pay for our sins, but this occurs within the probabilities of the events themselves. Divine retribution and/or divine tension is what we linear thinkers create for ourselves when we wish to atone—not for our sins, but because we weren't caught, because others didn't see us as the sinners we perceived ourselves to be. How linear our orientation is can be determined by the degree we feel punishment is necessary. Particularly in a linear-based society like ours, punishment is unnecessary because each individual invariably creates his or her own, generally in one of two forms: suicide or a desire to atone, to work especially diligently to better the self and the surrounding world in one way or another.

Simultaneity and Energy Collection

What does all this have to do with one form of energy supplying all our needs? Unless we collectors, creators, and users of any form of energy realize these are *simultaneously* the source of energy for other realities, we'll merely take this new and plentious form and use it in our wasteful machines and splash it around like so many drunken heathens at a hedonistic feast. Although most of us say we want unlimited energy—wealth, success, acceptance by all others— the truth is that linear philosophy doesn't prepare us for it in the least. When faced with the prospect of success, we tell ourselves we're being realistic and mature by tempering our enthusiasm with reminders of past failures. "I had _______ before, and look what happened. I lost it or nothing came of it." Rather than recognizing that *all* probabilities including the negative ones occur simultaneously with our successes and have already been dealt with, we create *new* negative probabilities and throw them out into the future to pop up like so many boulders suddenly appearing before us on a winding road, seemingly placed there by an unseen hand.

While the effects within one reality on a single individual can certainly be quite negative and even prevent that person from ever getting what she or he says they want most, such limited awareness regarding energy is even more destructive. If we look at the raw materials of fusion or fission versus their end-products, we can see that the instantaneous probabilities for such a system have little to offer. Regardless where we jump into the system, we must work with relatively incoherent forms; few of us would be comfortable with lumps of radioactive uranium and plutonium lying under our beds or the high heats necessary for fusion emanating from our ovens and furnaces. Until the system creates something we can use, such as electricity, all intermediate products are essentially useless as far as we're concerned. However this useless material, whether the result of an aborted reaction or waste from an incomplete one, functions as another reality's source.

Imagine two towns who agree to an exchange wherein each town decides what it will give to the other, similar to a grab-bag party. The first town bestows its most cherished works of art, but the second sends its garbage and raw sewage. Although our waste effects probabilities beyond our normal experience, we needn't travel far to see this phenomenon at work. When our waste is buried in the ground or dumped in the water, countless other natural realities are affected. Material considered too hot, acidic or radioactive for our needs suddenly becomes a source of heat, acid or radioactivity for fish, insect, microbial and mineral populations. Mythology is filled with tales about how the activities of different realities (gods and goddesses) created devastating or beneficial climatic changes on earth, depending on the nature of those deities involved. Those motivated by anger and hatred or other forms of incoherent energy invariably created desolate areas of extreme cold or heat; those motivated by love or pure coherent energy created bountiful harvests, pleasant climates, etc. It doesn't take too great an imagination to figure out what kinds of gods would dump radioactive waste on others. Our reality presently operates like the second town or an angry god: We take what we want and let others deal with our wastes. However if we merely filter coherent energy through incoherent sources, rather than structurally destroying those sources via compression or expansion, fusion or fission, and creating unusable waste, our role is quite different. Think of a sieve whose openings may be varied; if we want to hold back large particles we set the size of our openings so they're smaller than those substances. Now suppose our sieve is made out of the same material as that which we wish to filter, only in a different form—a rigid plastic filter for liquid plastic, one composed of ice to filter water, for example. We set our opening size and collect what we need. To be sure, if we filter out all the tiny particles, the remaining solution is composed of a greater concentration of larger ones. However, none of those particles nor the sieve have been destroyed or altered in any way. If others wish to filter out the middle-sized particles from the larger ones in the remaining solution, they can do so by increasing the size of the openings in the sieve. Still others may come along and collect the remaining undistorted components according to their needs.

Let's try another thought experiment. In the days to come, practice viewing events as simultaneous rather than linear. To do this, you may have to sharpen your ability to recognize the linear process of "first this, then that" before you can eliminate it. For example, when you sit with a loved one and the agonizing thought of separation suddenly flashes through your mind, it's because you're experiencing simultaneous probabilities. However, because the other is sitting blissfully at your side, you've consciously chosen to experience *that* probability. To take the worst which was simultaneously playing itself out somewhere else as a necessary evil state and perceive it as a sequel to what you're currently experiencing accomplishes nothing but unnecessary mental anguish. It's not necessary to experience the ante-matter state any more than it's necessary to ride beneath your car—unless, of course, you want to. The "good" events we experience have inherent in them a simultaneous "bad" probability somewhere. However, the bad does not precede or follow the good unless we create it that way. In other words, we can have our cake and eat it too.

Probabilities exist; whether they carry good or bad meaning is the choice of each individual. We've all heard about the thin line between love and hate. When viewed linearly that means we can never be sure when love will turn to hate. However when viewed rotationally, the hate need never be expressed because our awareness that it simultaneously exists and we choose *not* to experience it strengthens our love even more.

The Principle of Spontaneity and Being in Control

This Fifth Principle can also be called the principle of spontaneity. What makes pure energy work is that there's so much of it present at the same time—energy that spontaneously responds to the streaming effect the instant that first 7α pair moves. Until this happens there's flow without pressure, amperage without voltage. If we have a group of entities moving at the same velocity and aligned laterally versus linearly, each functions relatively independently. Again, think of troops marching uniformly beside one another versus marching in a line at ever-increasing rates. When the soldiers are all moving in the same direction in a lateral configuration essentially each functions as though alone. This is a critical point because so often in our energy awareness as well as our personal philosophies we believe that choosing to do something which is uniform to the beliefs or actions of others is similar to making *no* choice at all. How many people do you know whose only reason for not doing something is because "everybody's doing it"? Similarly, how often do young people in the throes of adolescence seek to assert their independence by not doing things they believe their parents and other adults want them to? In both cases to do what others do, even if it's the best thing, is viewed as inferior and to be avoided.

In addition to believing that a choice to be uniform is the same as no choice at all, we also often believe such a choice is somehow more permanent. The idea of being locked into a uniform pattern at best seems dull and boring; at worst, the idea is repulsive and frightening. Thus drivers respond much better to constantly changing speed limits, some responding to the actual physical act of acceleration and deceleration while others respond more to the emotional changes accompanying the act. In both cases the results are identical—we're seeking to affirm our variability, our right to change.

It's indeed a paradox that the uniform, simultaneous motion espoused by rotational physics and philosophy is often considered fixed and unmoving. Motion is the cornerstone of rotational theory; it maintains all is moving, all is constantly changing, and each person is capable of manifesting as much or as little of that change consciously as he or she chooses. In contrast, concepts such as chain reactions, conditioned responses, and inertia say humankind *only* changes in response to a lack of or excessive motion, like a vehicle having only two levers—one to start it moving and one to stop it. Because the vehicle isn't designed to *do* anything if it travels at a steady pace, it behooves us to create situations where the starting mechanism and brake can be used. Because we can only judge change in terms of starts and stops rather than the distance covered, we must also create the total system to support those beliefs. Rush hour traffic is a fine example of such a system at work. Many drivers complain about public transportation because it's too slow, they hate to wait; but many who prefer to fight traffic are quite literally addicted to the start-stop timing and have in fact geared their own physiology (mass), thoughts (energy-mass or just e-m), and emotions (energy) to it. If we view this in terms of product rather than process, we can say a reasonable goal of these people is to reach their destinations

- As quickly as possible.

- With the least amount of stress or tension.
- As economically as possible.

However, a quick survey of the thousands of single occupant cars daily jammed bumper-to-bumper in metropolitan areas all over the world suggests the actual goal is to be "in control". They say they drive because public transportation is too slow and they hate to wait, as though these were properties that don't exist in their daily start/stop, inch-by-inch journeys to and from work. The average commuter driving in traffic is involved in continuous chain reactions of one form or another. Petty irritations accumulate and explode as anger; one cigarette gives way to another and another; stomachs growl, then churn; people dash up the breakdown lane or run stop signs. Surely it's on the highways and particularly in the co-created traffic jams technological man loves so well that the majority daily reaffirm their belief in chain-reactions.

Can we use rotational theory to get out of this jam? Imagine yourself in a highly populated area known for its rush hour traffic, driving a vehicle having neither brake nor accelerator, traveling at a fixed 50 miles per hour. Now put these same restrictions on all vehicles and notice what happens: Are driving *skills* suddenly more important than driving *emotions?*

Negative emotions, such as fear and anger as opposed to spiritual energy, are the result of boredom as are all chain reactions. The only reason we need to get away from or accomplish a task as fast as possible is because that task is distasteful in one way or another. Whether it's dull or frightening, the effect is the same: We want it over as quickly as possible. However, for many, frightening experiences are sufficiently preferable to dull boring ones that we impart excitement by virtue of time. Even the most boring event such as driving to and from an unfulfilling job twice daily gains excitement if we build in a few physical and emotional start/stops and chain reactions on the highway.

Recognizing we create inertia or brakes in a system that naturally has none so we may then overcome it, and chain reactions that we may further accelerate or decelerate, also creates a bit of a dilemma. Removing inertia and exponential linear motion seems to take all the spirit and pizazz out of life. After all, much of what we technological humans describe as excitement is directly related to these phenomena—first their creation, then their maintenance and control. As you may have recognized in your thought experiment, driving the constant 50 mile an hour vehicle removed from the philosophical and mechanical constraints of inertia and chain reaction creates an entirely new set of needs. Suddenly maneuverability and flexibility become important, visibility is a prime concern and, above all, you must know where you're going. Compare your constant velocity driving problems to that of our chain reaction driver coming through the toll booth. The former requires skill; the latter, raw guts or total numbness.

One Form, One Philosophy

The concept of a universal, all-purpose form of energy is hardly new or unique and, not surprisingly, it's had its greatest attention as the philosophical concept of pure love. Christianity not only maintains that God is love and an energy essence, it's also filled with references to *in situ* energy production ("Consider the lilies of the field"), no resistance (opening hearts and minds to the wisdom and love of God), dangers of storage ("Store not up treasures on earth"), and adaptability (transfiguration). That the theory strikes a responsive chord is evidenced by the endurance of the religion. The decline of Christianity and other religions or philosophies is often due to the more superstitious, less logical leanings of particular philosophies at a particular time. In the past the most inquisitive gravitated towards various philosophies because these symbolically represented those intuitively known basic truths; but now many shy away because these philosophies have become increasingly structured and inflexible. Ideally philosophy and science work tandemly, one supporting and nurturing the other. Unfortunately technological time has increased so rapidly, philosophical time appears fixed and immobile by comparison. When Christ's disciples noted that love never ends, they vocalized one of the principles of energy, a concept capable of freeing an entire population and catapulting us into a world filled with new discoveries and experiences, free from the tension, stress, and fatigue created by variable start/stop energy in our technology and personal lives. But because they and we apply the principle to such a rigid, limited definition, it goes nowhere.

The fact that pure energy, like pure love, simply *is* in no way limits its flexibility. However, if we define love in terms of a set criteria—that someone must be or do something—the flexibility and purity is gone. As electrical potential even in its most impure, inefficient form flows from an outlet, what it can do is limited only by the imagination of the user. If the user is bound by pre-determined or fixed ideas, how that energy is used is bound by those restrictions from the beginning.

So great are our innate beliefs about pure energy that the inquisitive and strongly intuitive feel a great need, but also a great hesitancy, to test them. Those with classical, logical minds fare much worse in this endeavor because they're firmly grounded in this technology. Being grounded, a fairly popular term these days, simply means we believe something strongly enough to assign it the power of absolute truth *and* intuitive wisdom *relative to us*. If our beliefs about this reality are one and the same with our deepest, most profound intuitions about all that is, the system is quite stable, because it reflects all parts of our nature. However if our beliefs are singularly uniplanar and reflect only our views of this reality, being well-grounded is nothing more than being stuck within a highly limited mantle of beliefs. The father who feels threatened by the existence of his child's imaginary friend may consider himself a sane, logical and well-grounded adult. However, for centuries many cultures and religions have revered the ability of the most stable and attuned to communicate with all manner of beings and considered such individuals the epitome of what it means to be truly human. To them the child is more grounded multidimensionally than the parent can ever be without great change.

Principle Five is as much a philosophical concept as a scientific one and requires each individual probe deeply into his or her beliefs. Such a call can be threatening if we automatically assume those deep stirrings arise because something is wrong, because there's some demon aching to get out and wreak havoc with our orderly, predictable linear lives. Such isn't the case at all. As Christ proposed it, we know neither the hour nor the day; our recognition and acceptance of the meaning of pure energy and love as it supplies *all* our needs can indeed occur at any time because that intuitive awareness is always within us.

If energy in the most simple terms can do so much, do we create problems when we try to make complex energy-collecting devices? In the next chapter we'll discuss the relationship between specificity and waste in energy production.

9 | The Sixth Principle

The Sixth Principle: The more specific the energy-creation device, the more energy it wastes.

To understand the Sixth Principle, we must first talk a bit about probability, specifically uniplanar probability. Everything is probable: Nothing is absolute. Probabilities can be high (mathematically represented as close to 1.0), such as the probability of your drinking your cup of coffee; or they can be low (close to 0), such as the probability of your coffee suddenly disappearing the instant you pour it into your cup. We tend to ignore sure things like the sun coming up each morning, but are fascinated by the one-in-a-billion occurrence like someone winning a million dollar lottery. If the probability of an event occurring is very, very low—one in a trillion, for example—and the event occurs nonetheless, we often consider it a miracle.

Although theoretically probabilities can range from 0 to 1, these limits are never achieved. Remember the concept of the $_7\alpha$ probability mist described in *Primer?* There's never a place where a $_7\alpha$ can't be, but we can never be 100% certain where it *is*. This is a lot like turning a three-year-old loose in a playground filled with swings, slides, merry-go-rounds and see-saws, then turning your back and trying to guess where the child is at any given moment. It probably can't be done.

Most introductory discussions of probability begin with the familiar coin toss game of "Heads or Tails?". The probability of a head is 0.5 as is the probability of a tail. If the two probabilities are added together, $0.5 + 0.5 = 1.0$, that represents "all states of nature" as the mathematicians say. It also represents the numerical answer to the question, "What's the probability of tossing *either* a head *or* a tail?"

But isn't there a chance the coin will land on its side or roll across the table and fall into a container of acid and be dissolved? There certainly is. However, because the mathematical expression is designed to respond to the probabilities inherent in *only* heads *or* tails, it's an accurate description of these events. As long as we recognize that the probabilities reflected in any mathematical equation are only representative of the specific conditions defined and that these may not necessarily be the *only* conditions which may affect the system, we won't be misled.

For the sake of simplicity then, we're going to ignore all probable outcomes of our coin toss except a head or a tail. Mathematically, the probability of a head appearing is

$$P_h = 0.5 \text{ (or } 50\%)$$

and for a tail

$$P_t = 0.5 \text{ (or } 50\%)$$

The total probabilities for our coin landing either heads up or tails up is

$$P_h + P_t = 1.0$$

Knowing this, we can now figure the probability of a certain combination of results occurring in a series of tosses. For example, what's the probability of tossing two heads in a row? To calculate this we multiply probabilities:

$$(0.5)(0.5) = 0.25$$

For six heads or six tails in a row, the probability is

$$(0.5)(0.5)(0.5)(0.5)(0.5)(0.5) = (0.5)^6 = 0.015625$$

We can also demonstrate probabilities using what's called a tree diagram. A tree diagram of three tosses of a coin looks like this:

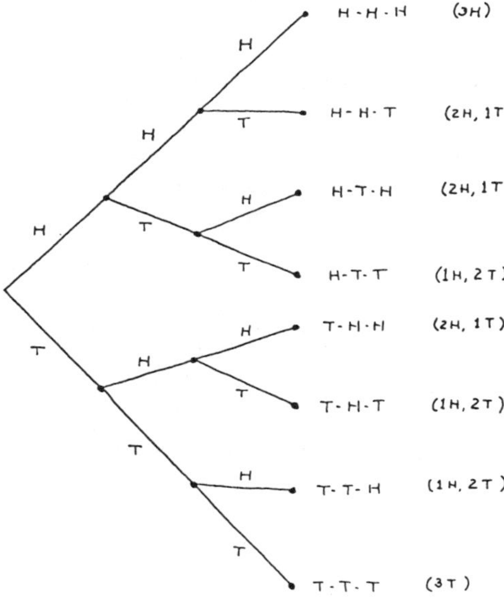

From it we can see there are four different possibilities by the end of the third toss:

- Three heads.
- Three tails.
- Two heads and one tail.
- One head and two tails.

Furthermore, our tree tells us how many times we can expect each combination to occur as we flip our coin:

- Three heads will occur one time out of eight.
- Three tails will occur one time out of eight.
- Two heads and one tail will occur three times out of eight.
- Two tails and one head will occur three times out of eight.

Using these data, we can answer many more questions about probability. For example:

- What's the probability of tossing *exactly* two heads in three flips of a coin?
  Condition 3 (two heads, one tail) occurs three times out of eight or 3/8 = 0.375
- What's the probability of tossing two heads *or more?*
  Condition 3 (two heads and one tail) + Condition 1 (three heads) = 3/8 + 1/8 = 4/8 = 0.5

Why do we need to understand probabilities? Because much of what we consider possible for the future comes from our observation of either what is now or was in the past. To be sure, if we examine a pair of dice, we can discern the probabilities of any throw without knowing who did what in any past crap game. For instance, we can see that each face of any one die has a one-in-six probability of showing up. Therefore, the probability of throwing "snake eyes" or two ones is

$$\left(\frac{1}{6}\right)\left(\frac{1}{6}\right) = \frac{1}{36} = 0.028$$

However, when we look at something more complex like energy production or open heart surgery, all the possibilities aren't so obvious. In such cases we find ourselves noting "When Dr. X used this procedure, he got these results." When we note Drs. Y, Z, A, B and C got similar results, we feel more comfortable with the procedure because we have some idea what results it produces. In time we even feel confident predicting what will happen when that procedure is used.

The past probabilities come from the belief that once a certain pattern has been established in the past, that pattern can be expected to continue in the future—which itself is probabilistic, but that fact is usually ignored. For example, many people assume milk is good for cats when in reality many cats are allergic to it. When cats were first domesticated to control rodents, farmers fed them milk because it was readily available—not because it was nutritious. However, the past pattern lingers and will continue until enough cat owners experience problems to create a more representative set of probabilities.

Suppose you're due at work at 8:00 a.m. and over the past five years your arrival times have varied as follows:

Time of Arrival	Percent of Time
Before 7:50	3
7:50–7:55	7
7:55–7:58	10
7:58–8:00	30
8:00–8:02	30
8:02–8:05	10
8:05–8:10	7
After 8:10	3
	100

We could take this past information and construct the well-known, bell-shaped curve:

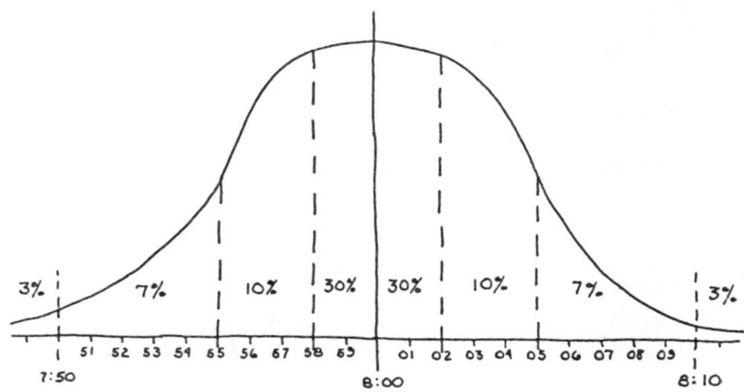

If someone asks your boss, "What's the probability of Harry being more than five minutes late?" he or she could look at the graph and answer "10%", adding your 7% arrival time between 8:05 and 8:10 and your 3% arrival time after 8:10.

Because we know we tend to think in terms of probabilities, obviously probabilities play an important role in *how* we think. Unfortunately, many times we lose sight of the fact that what we may now consider absolute truth was at one time, and still is, merely a probability. Instead of accepting Drs. X, Y, Z, A, B and C's results as the more probable, we say they are the *only* results. In such a way probability gives way to specificity, the former's exact opposite, and all sorts of problems arise.

Specificity

What are some of the problems associated with specificity? First, a belief in specificity as *better* often arises with an associated belief that the more specific or defined a *process* is the more control we have over it. However the desire to control the process of energy creation via specificity, like the desire to control love, is extremely time-consuming and inefficient. Consider the following: *All* definitions of what energy/love does, should do, or can do are, in reality limit its creative process. Specificity invariably leads to process orientation; having defined our little empires, it behooves us to create rules and regulations to denote our power and prove we're in control—whether our specific empire involves producing energy, generating foreign policy, training raw recruits, manufacturing contact lenses, or organizing the quilt raffle at the church bazaar. When we say every electrical circuit *must* have pressure (voltage), flow (amperage), and resistance,* we incorporate those characteristics into its process as well as its use. We can't imagine creating energy without these parameters; we can't imagine utilizing or altering energy without altering one or more of those parameters to create a different form. Similarly we can't imagine foreign policy unassociated with defense, training recruits without "discipline", or someone other than Mary Ellen handling that quilt raffle.

If we think of love as the internal energy we each generate, we can view its action exactly the same way we view the energy form of the singular $_7\alpha$ and its interactions. That being the case, the spiral $_7\alpha$ with its spin, linear velocity, or combinations thereof, is also totally representative of the actions of all greater rotations— atoms, humans, galaxies.

But how can we get from what we can see, or what is real to us, to the $_7\alpha$ level which is too small to perceive? Look at a table

---

*With your knowledge of rotational physics/philosophy, can you decide which components correspond with body, mind, spirit?

in your room. If you're like most people, you see a piece of furniture with objects on it. To you it may look like this:

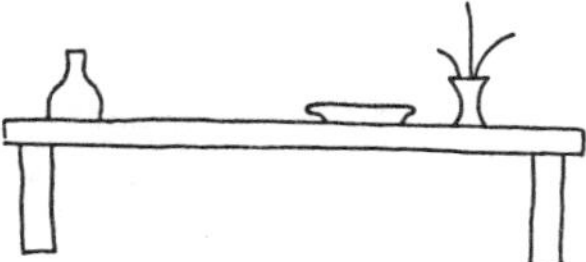

However, rotationally it appears

like a bunch of bed springs. These are the so-called "energy fields" surrounding and enveloping the object itself, creating its probability mist. Where the majority assume sufficient V_S and/or V_L for perception to occur determines the area of maximum "realness". For example, in very dim light you might not find the edges of the table or the objects on it quite so real. Perception-altering substances such as drugs or alcohol could similarly shift your idea of real, causing you to run into the table because you didn't see it, or reach for something on it that wasn't really there.

Euclidian vs. Rotational Thinking

That being the case, we can see how we're actually governed by two criteria when determining real: traditional uniplanar or Euclidian thinking and rotational or multiplanar thinking.

Actually, the difference between Euclidian and rotational mathematics or thinking isn't so much a matter of linearity as a matter of time. Euclidian thinking concerns itself with here-to-there phenomena—an apple dropping, a train moving, a bridge spanning two points. As such, it's a point-in-time evaluation denoting either what's happening at a single point or the change occurring between points. Rotational theory describes the *total probable identity* of an object or system and, as such, involves all or no time—depending on your point of view. We can compare Euclidian theory and its results to taking, developing, printing and framing a photo of a child. A comparable picture rotationally results in multiple, simultaneously distinct and perfectly clear images. Although there isn't one single *tangible* or absolute picture we can put our hands on, we do have access to an infinite number of images to which we can impart sound, light, emotion at will. So while on the one hand we can say the result of the Euclidian process is more real and "useful", we also recognize it's extremely limiting and represents only a single probability out of the infinite number available.

How does this fit in with the Sixth Principle? Like this: As we become more specific, more Euclidian, about anything, we lose flexibility and thereby tend to waste or even ignore other options. If we want to know the *total* capability of any system, we must view it rotationally for only that form takes into account the simultaneous existence of *all* probabilities; if we want to define the capability or potential of a system at one point in time, we apply Euclidian principles. However, the problem arises when linear Euclidian results are interpreted to mean that's *all* there is. Imagine a negative balance in your checkbook: Euclidian thinkers would take that point-in-time calculation and use it to predict events in the future or in the past. In other words, they use the point-in-time calculations of the present to predict *process* in some other time. So the single manifested probability, the negative balance, becomes synonymous with no food or shelter, or credit-rating ruination some time in the future, and perhaps poor planning or frivolousness in the past. This is akin to taking the picture of your child and imagining how that *picture* will look in 10 years, or looked ten years ago. That is, instead of imagining how your *child* will look in 10 years, you concentrate on how the picture itself will change. Will it fade or wrinkle? Will the frame fall apart? If your initial purpose was to study the effects of time on photographic paper and framing materials, such emphasis would be valid. However, if your purpose was to capture a single image of your child at a single instant in time, such evaluations regarding process have little or no meaning at all.

Reversing the Probability Tree

Similarly, when we take Euclidian evaluations of singular point-in-time events in fission or fusion reactions and use them to predict future or past process, our interpretation of what is really going on could be quite distorted or even non-existent. Like the parent who carries a strong mental image of a child based on fixed mass appearance rather than a less distinct but far more powerful multiple love energy "image" often finds it difficult to recognize mass changes in the child, so scientists who use mounds of point-in-time data to predict reactions occurring outside that time frame can become equally confused. This type of thinking takes a single point-in-time calculation, selecting one of a number of infinite probabilities available at that time, then uses a series of such singular highly specific experiences to predict a singular event in the future. Probabilities so used aim to reverse the probability tree:

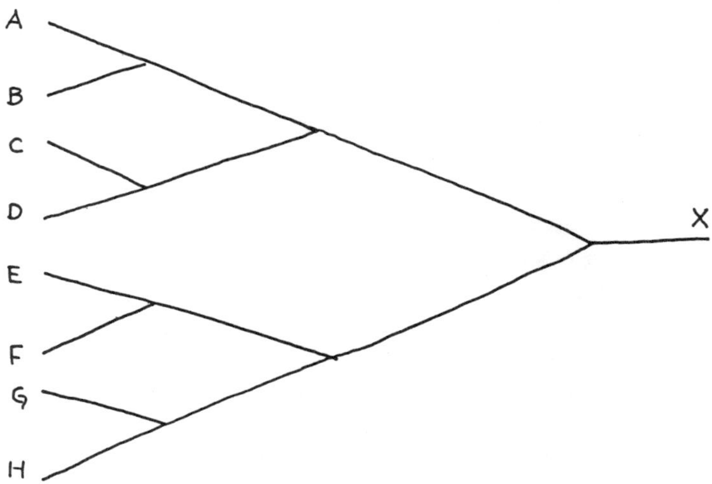

By totally defining a series of singular point-in-time probabilities (A, B, C....H) the idea is to predict an absolute event, X. Furthermore, it's believed that the more probabilities we can define, the more accurately we can define X. For example, parents with fixed views of their son may see him as

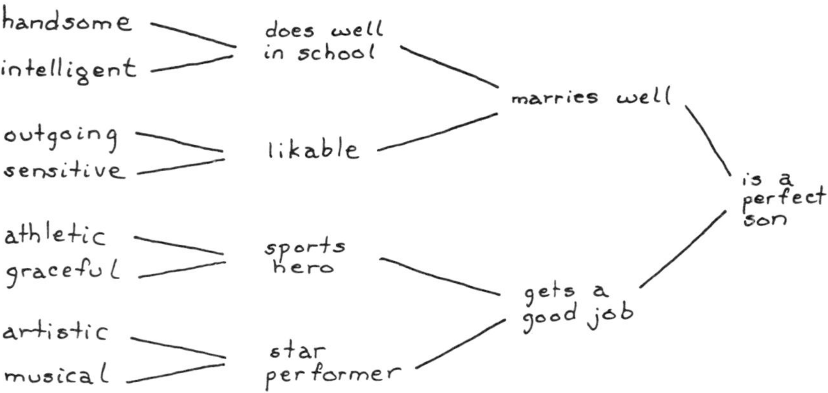

When the child was very young they applied their definitions of handsome and intelligent to him and eliminated all other probabilities. For example, their definition of handsome might include the boy's being bigger than his peers or having dark hair and blue eyes like his father. If his mother were sensitive about her large nose, much of the son's physical attractiveness to her may center around his delicate button-nose.

Having defined these point-in-time qualities, the parents then use them to predict future events in their son's life. Because he's athletic and graceful according to their definition, they expect him to excel in sports. If he uses each one of his qualities as they expect, they assume he'll experience certain specific events in the future. Those events, in turn, will lead to other events, the ultimate result being a "perfect" son.

From this we can see that the greater the number of defined qualities the parents attribute to the child, the more intermediate steps there are between those initiating qualities and the desired perfect goal. Surely we've all seen or read dramas involving parent-child conflicts in which the parent maintains the child has everything going for him or her and wastes it, whereas the child maintains nothing he or she ever does is enough to please the parent. So in fact, the more we depend on highly defined point-in-time input to predict future output, the more highly defined that future output must be for us even to recognize its presence at all. Thus for parents whose ideas of what their child is or isn't, can or can't do, are so rigid they can only envision one result, when another result occurs (their child becomes a monk instead of a general), they may fail to acknowledge either the result or the child, or both.

If we take our tree and replace the childhood qualities with the basic ingredients in an energy releasing reaction, we can see the same sequence of events:

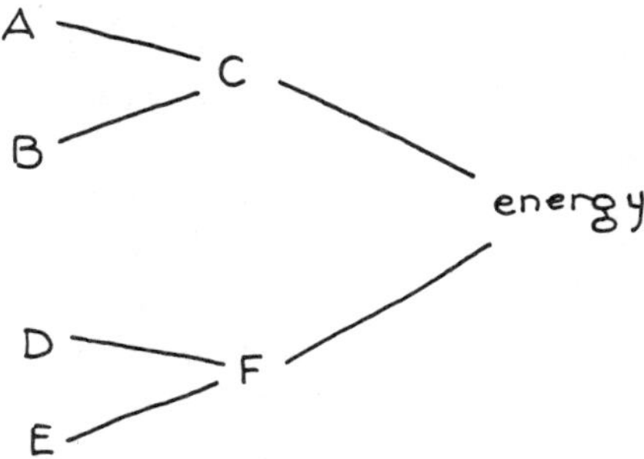

By definition, when A and B react they produce C which then combines with F, the result of D and E, to produce energy. Furthermore, the reverse is also held to be true: If energy is released, it must have come from the breakdown of C and F which were composed of A, B, D and E. In other words, the result is used to prove the validity of the point-in-time initiators and vice versa.

Although this may be true, it doesn't *have* to be, and may not be the *only* truth. Imagine a perfect forty-year-old son. Perhaps your image is of a white, Anglo-Saxon systems engineer, father of two, member of the church council. Or perhaps your perfect son is a black physician, or a homosexual graphic designer, or a paraplegic priest. The fact that *all* of these are equally valid forms of the perfect son is a major difference between rotational and Euclidian thinking.

Rotational thinking eliminates all the intermediate process. In energy production, our goal is energy. How do we get energy out? We put energy in. We get out what we put in: Those who live by the sword are indeed doomed to perish by it. If we recognize 50 probable energy forms we get out, then we must create a process whereby we *must* put 50 probable forms in. If as parents our goal is to love our children, that is all we need put into the reaction. Regardless what they or we *do*, the love will always be there.

Creation and Filtration

As we discuss the concept of specificity we must be careful about our use of certain terms. We've already noted how people often use the phrase "energy *creation*" when in fact they're really referring to energy *filtration;* considering the complexity of the processes, this is an easy mistake to make. However, because we'll be discussing energy creation or the conversion of mass to energy *here* as well as filtration and collection of the already existing energy, it's an important distinction to make. It's also an important philosophical bridge to cross as well. So often we want to believe we create energy because creation implies control.

The fallacy of such thinking is demonstrated throughout the Bible, Koran, Book of Mormon and in multitudes of other works: Those who believe creation gives them control are invariably proven wrong. Think of the fate of rulers who believed they created a particular government and therefore had the right to control all within their creation. Or think of the problems Pappa Guipetto, Pygmalion and Dr. Frankenstein had with their creations. In all these instances, creation wasn't only intimately related to control, it was intimately related to the control of a *specific* form. The reason the creators couldn't cope with the creations was because the latter didn't turn out in the expected way. Compare this to the creator seeing him or herself as a means whereby the creation can manifest its own potential in its own way. Here the creator functions more like a filter, providing guidance, but not *irreversibly* altering the basic form.

The *only* creation over which we have total control is the self and even here our control is strictly limited to change. The self, like the $_7\alpha$ spanning all that is, is an integrated whole; although we may choose to change, that in no way means we lose the potential to manifest in some other form in some other time and/or space.

Specific creation says you look at a piece of cloth and see only a black coat with brass buttons and narrow lapels, the *exact* garment you need for a particular theatre engagement one particular winter evening. The concept of creating change/filtration says you look at the cloth and see a potential coat, suit, skirt, table covering or drape, depending on how you choose to use it. In both cases you're taking your likes and dislikes, your feelings and ideas, and giving them form—you're converting energy to mass. However, the specific creation approach only results in one form, the evening coat, whereas the approach preserving change results in many variations.

What philosophers and writers of the countless stories like *Frankenstein* and *Pygmalion* discover is that our creation by definition must invariably come from *beyond* here in that we're taking pure, inspirational, spiritual energy, changing it to the energy-mass form of thought, and then mass. In order to then control it, we must know ourselves on all those levels. What usually happens is that we willingly acknowledge our physical and mental input but not our spiritual energy. The only reason people create things they perceive as greater than themselves is simply because they don't know themselves. If they intuitively create something that manifests qualities they believe are beyond them, they're invariably threatened by their creation. Whether they succeed in destroying their creation or vice versa depends on what they choose to prove. If they want to prove it's impossible for the creation to exceed the creator, they either destroy it or live in peace with it. If they wish to prove the creation exceeds their own ability, they permit it to destroy them. Again, history and literature are filled with numerous examples of this phenomenon.

The spiritual input is most commonly disregarded, particularly if it contains doubts, and yet it is, as the self-fulfilling prophecy (SFP) has shown many times, the most powerful component of all. There is yet to be a structure built by creators who doubted one iota of their skills that remains or will remain standing. Depending on the builder and his doubts, the structure may collapse during construction or after the builder is gone from the area or even this life—*but it will collapse*. You may counter by saying all structures will collapse, but we suggest you ask yourself: "Why?" If there are no external forces or if the external forces exactly balance the internal, why should a structure collapse? There's no reason for such to occur if it's at one with itself. The problem is: How can anything be at one if its creator doesn't believe in it? The effect of that belief on the energy and e-m components is bound to affect the mass form. Think of placing a large electro-magnet near a steel building. In spite of the great mass of the building, it's constantly straining toward the magnet and against the foundation. Similarly, the presence of doubt—the need to control—creates its own SFP, a huge magnetic field which invariably destroys our own creations.

Another way to view this concept of energy creation is via the example of love. We only create love in ourselves; we can't create love in another. The singular $_7\alpha$ may convert to 100% linearity or spin, or any form between, but it can't *make* another $_7\alpha$ do likewise; nor can it experience any change in itself without affecting all others. Our love for the self, our form if you will, is the *only* change we can create. The more we love or energize the self, the more we affect others.

Specificity and Purpose

Specificity implies limitation. Does this mean a strong sense of purpose is limiting? No: An awareness of purpose is essential to our feelings of success. However, a highly defined idea of *how* to reach that purpose, its process, is detrimental. Whenever anything is rigidly defined, the system immediately becomes wasteful and inefficient. Like the person who views eggs as only coming from chickens and only for making cakes, so those who view potential or electrical energy as arising from a specific process for use only in those appliances producing friction or resistance are equally limited.

Specificity of either process or result also implies potential limits of amount. The more specific we perceive the process to be, the more we realize a breakdown in that process will destroy our singular supply. The more specific we view the product to be, the more we realize a change in whatever uses that product will render it useless. Again, this is most easily understood if we view energy in terms of love.

Process specificity is like those who have highly defined ideas of what others must do to make them feel loved. If certain words or acts aren't forthcoming, they believe themselves unloved. The fallacy behind this thinking is two-fold. First, they're bogged down in process and secondly they expect another's actions to confer love upon them. Just as those who create one inefficient appliance after another to use an inefficient energy form are never satisfied, so those who depend on others to define and manifest love for them are never fully satisfied because the basic premise is unsound.

The concept of specificity isn't new. Were humankind to apply specificity to the concept of energy on the smallest level or result rather than its creative process, all would be well. As it is, sloppy generalization is applied such that the Fifth Principle (All needs can be fulfilled with one form of energy.) is interpreted to mean a different form for every different task—creating multiple configurations and higher-level rotations, each with a simultaneous, most rigid process to collect that form and a highly specialized machine to use it. Far better to have one process to filter and subsequently collect all forms.

The Fifth Principle, then, does indeed tell us love hopes, believes, and endures all things; love never ends; love is not arrogant or rude. In other words it's always there when we need it, however much we need, and without waste or pollution. The Sixth Principle notes that the more we define our love for others or that which we receive, the more waste and distortion of that love occurs.

Specificity and Resistance

Let's examine one last energy-love parallel before we go on to the Seventh Principle. There's a prevalent belief that good things, including love, don't come easily. Many of us feel a certain force such as long hours, hard work, or courtship must be applied to any system to overcome resistance or friction. As we would expect, this belief carries over into physical mechanics as well as electrical theory. Ohm's law states that voltage or electrical pressure (E) is equal to amperage or electrical flow (I) times resistance (R):

$$E = IR$$

Therefore, the less resistance there is, the greater the flow. In rotational energy production we want greater flow rather than more force. Pressure (E) is simply force per unit area; that being the case, Ohm's law tells us we can achieve greater flow by reducing resistance as well as by increasing force. So, if we rearrange the equation to express flow,

$$I = \frac{E}{R}$$

and let the resistance approach zero,

$$I = \frac{E}{0}$$

then

$$I = \infty$$

and the flow becomes infinite.

Whereas linear approaches often concentrate on creating more force, rotational theory's primary consideration is always: How can we decrease resistance? As soon as we increase force, we create waste and lose power. Power is work or energy expended per unit time, and work is force through a distance. Work, then, is a moving force: flow.

Although we recognize that both Euclidian and Newtonian physics describes work (W) as a function of force (F) and distance (d),

$$W = F \times d$$

we often fail to notice that this elegant equation allows no room for friction or resistance. If we want such impedances we must *add them* to the system. This tells us something very important: The fact that the relationship expressed between work (multidimensional energy flow) and power (unidimensional flow per unit time) is valid *without* resistance, friction, impedance, drag or waste, says it's possible to create systems where such hindrances don't occur.

The difference between emphasizing force rather than flow is the difference between expressing anger and love, tightness and openness, tension and freedom, lies and honesty. In the next day or week, practice being open and accepting of others and their ideas rather than closed and rejecting. This doesn't mean you must suddenly believe as others do; it simply means you accept that people are entitled to their own beliefs as you are entitled to yours. It's the Christian's Golden Rule and the flower child's "Let it flow."

What happens when the resistance is suddenly removed from a system geared to produce force only against a fixed resistance? Think of water pounding against the closed gates of a dam which are suddenly opened: decreased force, increased flow. It's the very basis of the oriental martial arts which creates some of the most powerful fighters embodied in the most gentle, respectful, and diminuitive of all people. The attacker or force comes expecting resistance; when there is none, when the intended victim suddenly moves aside, the attacker's force is converted to flow causing him to lose his balance and fall.

Blessed (powerful) are the meek (non-resistant).

When we decrease our resistance, we increase our power. However, we can *only* decrease our own resistance, not another's;

nor can we "over-power" *them*. Decreased resistance, openness, honesty can only be accomplished on an individual basis. Unfortunately we often function as though a decrease in our resistance is an open invitation to some outside force to come and overwhelm us. Rather than seeing a decrease in our resistance, an increase in our honesty and willingness to express love as a means of increasing the flow of our own love or energy, we see it as an invitation to destruction by others.

What do you suppose would happen if one morning one of the two super*powers*, the U.S. or the U.S.S.R., announced that by the end of the day all their weapons would be dismantled, their military forces disbanded, their covert operations stopped, and all classified information made public? Would the other power then come to destroy it? Rotational theory strongly suggests such an aggressive reaction would be *impossible*.

If specificity is such a wasteful orientation, it would seem that the idea of storing energy won't fare too well rotationally either. In our next chapter we'll discuss storage in detail and see how, and if, it fits into our theory.

10 | The Seventh Principle

The Seventh Principle: Whenever energy is stored, more energy is lost than gained.

Those of us in Western society have extended the concept of storage to ludicrous limits. We amass money in all forms—savings accounts, investments in real estate, stocks and bonds. If we can afford them, we accumulate homes, cars, clothing or jewelry in staggering amounts. We often say we're saving for a rainy day even though rainy day savings tend to create no-win situations: If the rainy day occurs, the self-fulfilling prophecy says there's a good chance our saving for it helped create it; if it doesn't, the money sits in a bank of no use to anyone, except maybe the bank.

Doing away with our beliefs about storage isn't easy. Let's do a little thought experiment. Imagine that all you have in the world is $10,000 in a bank account. Our rather unusual bank announces that for one day its patrons may either withdraw money from or may make deposits to *your* account. How would you feel? Would you be frightened and angry, or excited and happy? Most people would be uncomfortable at best. Or think of the idea of love flowing unchecked from and to you; this is also unnerving to many people. In both cases the predominant feeling is one of vulnerability and powerlessness because in linear terms we associate power with storage. We believe there's no pressure, no force, without storage. Although nothing could be farther from what is, look around you: Practically everything *man-made* in this technological world reinforces the concept of storage. Is it any wonder it's so difficult to decrease our resistance sufficiently to permit the free flow of something so valuable as our love?

According to Euclidian thinking, the more we wish to express love, the more force and power it must have. However, by its very nature love isn't linear; it's a function of flow, not pressure or force. Therefore, linear evidence of true or pure love will not, and can never be present. Love is energy, not force; energy is flow, not force. Because there's no flow in storage, there can be *no* energy. We may store force (like stockpiling weapons), but force has no power, no meaning unless it encounters something willing to yield to flow in response to it. In such a case the force is dissipated and becomes nonexistent, while the one willing to flow is the one ultimately manifesting the energy.

Storage creates pressure. Although pressure or force may be considered the "source" of flow, Ohm's law has already shown us that as resistance approaches zero, flow becomes infinite,

$$I = \frac{E}{R} = \frac{E}{0} = \infty$$

I = flow (amperes)
E = force (volts)
R = resistance (ohms)

and the value of E may approach zero without affecting the infinite nature of I. In other words, if there's no resistance, *that* becomes the determining factor of flow rather than the force.

Over 90% of the waste occurring in energy production and its use results from storage. Technological society has the erroneous idea that conservation refers to the use of the same excessive amount of energy for shorter periods rather than the continual use of lesser amounts. It's the turning on and off, the startups and shutdowns that create the waste. When we apply the principles of conservation correctly to wildlife, we don't go out in spurts and hunt over-populating species, then allow them to over-populate again; we maintain a continuous process to create and maintain a stable population.

Continuous Flow vs. On/Off Systems

If we speak of uninterrupted flow as the most efficient way to use energy, isn't it wasteful to have energy flow when or where we *don't* need it? Let's look at this another way: The average household is always consuming some form of energy—clocks run constantly, indicator lights glow. Instead of viewing the source of energy to the house as a reservoir and each switch to each light or appliance as a dam, let's apply our theory of flow and gates.

Naturally if we have any system using gates, whether water, energy, or love, the more gates open, the greater the flow. And open gates create a streaming effect; if all gates are closed, there's no streaming effect and therefore no flow. If there's no flow, there's no motion or energy. In our average, imaginary household in the year 2000 we have a nonspecific collector, most likely sitting in the yard or replacing the roof. It's a closed system like modern solar hot water heaters and uses the principle of feedback; the more gates open within the house, the more open in the collector. Thus the flow is *always* continuous, there's *always* enough energy for what we're doing, but *never* storage or waste.

When we rely upon on/off systems based on pressure or force, we're severely limited by our own creation. The greater the pressure or force inherent in a system of production or control, the greater the mass to be overcome to put the system in motion and to use. Like a large raging beast capable of moving mountains, kept tranquilized, chained, and imprisoned within an electrified cage underground, storing current sources of electrical pressure to do work sometime in the future is an arduous, wasteful and dangerous process. Far better to have billions of tiny, friendly and harmless ants eagerly working around the clock.

As long as there's no flow, there's no resistance. The toll gate closed on an empty turnpike offers no resistance to traffic flow. If flow exists, obviously the more gates open, the faster, more efficient the system. The streaming effect tells us that if some gates are open and others closed, all flow is drawn to the open gates. The resultant flow always has the same velocity, but the number of gates opened or closed determines the amount or volume of the flow.

Storage and Flow in Human Interactions

We may also see the principles of flow and gates at work in human interactions and emotions. Compare the three forms of love—erotic (romantic), filial (friendship), and agapé (spiritual or godly love)—to body, mind, and spirit (mass, e-m, and energy) and also to banks of gates in an energy collector. How many and which gates we open determines the amount of flow in terms of influence. For example, if we have nothing in common with another person, only one filial gate may open and only the most *narrow* stream of energy goes out to that person. You might vaguely take note of that person when he or she asks the correct time, but then immediately ignore them. If you dislike or even hate another, all gates of filios and agapé may be affected. If your hatred is so great it "inspires" (in-spiritualizes) you to physical acts, the eros gates are also affected. We say affected because often what occurs is an internal struggle to keep those gates closed. However because the flow of ill-will or incoherent energy isn't stopped, pressure is created. If we continue to permit the flow of more negative feelings, they build up and keep spilling over until they find a system of gates which *will* yield. For many, this manifests in either outwardly-directed acts of physical violence or self-directed incoherent energy which then results in discomfort, disease, or even death.

We can always give totally without losing control because it's totally within our control how many and which gates to open. In such a way there's no resistance, no storage, and therefore no pressure. The interesting and quite comforting thing about the human psyche is that, whether we choose to consciously acknowledge it or not, we're *always* giving 100%. Whatever we're giving and whatever we do with it always represents exactly the total amount of energy we wish to flow in that situation. This means that any *effects*, whether positive or negative, aren't the result of the *amount* of flow, but how we choose to channel it.

In the days ahead, practice becoming aware of *what* gates you open and which ones you hold closed. Don't worry about why; just become aware that there are certain energy channels you wish to share with others and some you don't, and that the fact that certain channels are closed doesn't decrease the rate of energy flow. Once we can acknowledge this, we can be free of the crushing guilt that's often associated with the erroneous belief that true (100%) love for all persons and things necessitates opening all gates. True love means we channel all our energy-love through the gates we choose—no more, no less—being fully cognizant of the fact we get back what we put out. Our most valid indicator of what gates are open is the other's response. If different gates are functioning—you channel via filios, and another is more interested in eros, for example—you'll be aware of a *difference*, an incoherency of flow and the resultant pressure or force. ("The two of us just aren't on the same wavelength!") Then it becomes a matter of whether the two participants wish to maintain this incoherent, pressurized exchange or align themselves to one channel or another.

Storage Negates Purpose

Another problem inherent in storage is that, by definition, it negates purpose. Unless our purpose is to create storage (to study its effects, for example), there's no reason for it. One modern religion requires its followers keep one year's supply of food on hand: What is the purpose of this? We suppose it has something to do with survival in case of disaster, but doesn't that imply a certain lack of faith both in themselves and their God? Also you can imagine what could happen to a family that has a year's supply of food when those around them have none. Indeed, they could start worrying that "thieves may break in and steal."

If we know what we're creating energy for, there's no reason to store it. If we want light at point C, we simply create energy at point C; to create it at A, store it at B, and flow it to C is wasteful and has no purpose. Although it's accepted that within most electricity-generating systems, voltage or force is created, it's also acknowledged that the flow or amperage actually produces the results. The only function of the pressure is to permit *sudden* initiation and cessation of flow and the accompanying emotional gratification we get from being in control of an on/off device. Most rivers flow freely without pressure. It's only when we try to control or store that flow that we create problems. How would you feel living downstream of the Grand Coulee Dam? Comfortable?

Put yourself in the place of a $_7\alpha$ within a lump of fissionable material inside a reactor. You know you want to illuminate a lamp in a house at 137 Main Street at 6 p.m. The process standing between you and your goal seems insurmountable. Like a passenger on a 10-mile trip being asked to shift from car, to bus, to train, to bus again, each accompanied by periods of frenzied, confused activity, followed by long waits in terminals and sidings, eventually you come to wish you had walked. Interestingly, the more impatient we are the stronger our intuitive awareness of natural flow. Imagine yourself driving behind a loaded logging truck going up a five-mile steep, winding road when you're already five minutes late for an important meeting at the top of the hill. Do you feel yourself becoming tense? Does your (blood) pressure start to rise? Impatience is an emotion, an energy form that needs some place to go. When we tighten up, close gates and increase resistance, pressure must increase. The energy, the flow, the emotion is there; we can't make it disappear. If it can't pass or flow through because of our resistance, it must manifest as pressure.

Learning to Use Energy Flow

Put yourself behind the logging truck and deliberately tense up; feel what it does to your body, mind and spirit. Now relax, imagine millions of tiny holes in yourself and let the feeling dissipate and flow gently away. There's no magic to it; it's how biofeedback works. It's a simple survival skill anyone can teach him or herself.

Learning to use energy flow within yourself is also a lot like learning to drive a brakeless car whose accelerator goes from zero to infinite speeds, similar to "dodgem" or "bumper" cars in an amusement park. In this thought experiment, our cars are frictionless; when you slow down, you slow down. The vehicle responds *only* to the amount of gasoline you provide; steep inclines, up or down, have no effect. Only more gas makes you go faster, any less gas slows you down. Compare the way you drive this vehicle to one having brakes. At first you may feel extremely vulnerable because your belief in the outside forces acting on your car and against your will is so strong. However, once you realize there are *no* outside forces, you quickly learn to handle the new car with ease.

Now let's take our brakeless vehicle and approach a barrier; mentally test your reactions. Add the extra dimension that the car wants to keep moving although not at a constant velocity. How do you solve this problem? What are your observations under these circumstances?

Both human physiology and the psyche function on the principle of flow. The reason acupuncture works for those who believe its purpose (better health) is because it helps to restore the psychic flow to physical portions of the body. Essentially, the metallic needles serve as reverse antennae, drawing energy to them rather than radiating energy outwardly. The human system wants to be at one, it seeks to manifest its energy and *it always does*. If we "trap" energy within us because we believe its effects will be negative if we allow it to flow through, then that energy must manifest inwardly—often with devastating results. For example, if a person is afraid to grow or otherwise change mentally or spiritually, then "growth" will occur in other ways. If the excess energy manifests as a mass change in the body, the result could be exactly that—a mass or tumor, benign or malignant.

Thus whether we like it or not, whether we choose to consciously acknowledge it or not, we supply ourselves and all others with evidence of our own energy levels. Watch what happens the next time you're in the company of someone who regards you as superior to them in one aspect or another—your daughter's new boyfriend, your new assistant making his/her first presentation, for example. You may detect fidgeting, random movement or even displacement activity such as eating, drinking, watching TV, foot tapping. Things like nervous laughter or coughing fool no one, least of all the self: Energy created in response to purpose or need is always expressed. Obviously the behavior others display when they feel inferior to, or uneasy with, us is probably quite comparable to our own behavior when we find ourselves in similar situations. That being the case, we can see how

- Storing energy is useless.
- Trying to channel it to some other purpose is wasteful and creates incoherency within the system.

If we create emotional energy with the purpose of responding to someone else's statement ("Joe, I think your plan stinks!"), but surpress that response, the flow has already been created and must manifest itself. Often the highly linear-energy form of emotional purpose becomes diverted to physical activity, the so-called nervous energy—pacing, chain smoking, nail biting—which can wreak havoc. Without the spin component, the energy passes through (See *Primer*.) and does not stimulate the ancillary systems which coordinate the physiological functions. Depending on the individual, one common result is blind rage accompanied by unchecked muscle activity; in other words, physical violence and great anger. More commonly in our "civilized" society, however, the energy is suppressed and the body forced to deal with the increased pressure. Many an individual who believed displaying emotion and thereby permitting energy to flow through wasn't proper or manly behavior wound up living with the physical manifestations of their beliefs: high blood pressure, stroke, heart disease. That's a pretty high price to pay.

So we may say that human physiology and psyche are natural systems of flow; the engine is always running and the brakes are no more than less engine. All energy created *will* be manifested in one form or another.

Storage, Breaking Mechanisms, and Control

Obviously if something is stored, we must be able to turn it on and off; therefore we can't have storage without some sort of starting and stopping mechanism. The concept of braking mechanisms as a means of control is so widespread that 100% of the engineered, legal, and medical systems in use function with that as their goal. For example, let's take a look at prejudice and discrimination. Those most vocal, most rabidly prejudiced against any group, species, or idea are invariably those whose contact with these others represents a 10% or less commitment to full knowledge of that group. Therefore any anti-discriminatory legislation is aimed to control or put a brake on behavior which is representative of 10% or less of the population. Such legislation is based on "an ounce of prevention" and a belief that the worst always happens. So the fact that 10% of the population is prejudiced against another 10% creates legislation applying to everybody, but totally unnecessary or mean-ingless to 80% of the total population. Only the 10% who unequiv-ocally oppose a certain group or idea and those who unequivocally espouse it, feel they've "won" or "lost", depending on the wording of the legislation. The majority continue as before, dealing on an individual and daily basis, being for something today they were against yesterday. This doesn't mean these people are stupid, in-different or capricious. It does mean they intuitively recognize there are no such things as absolutes and that specifics are limiting. Ba-sically any law which attempts to force a fixed behavior on others is merely our way of saying we're afraid of flow, we're afraid of change. If we can keep others from changing, from being different from us, then maybe we won't have to change ourselves.

If flow is the purpose, why is force—either a positive accel-erating or negative braking force—necessary? If a string of aligned electrons can initiate the streaming effect, why must we pile up trillions of them awaiting our beckoning call? Like anything else, our need to control is always greatest regarding that which we know, love, or trust the least. The individuals least concerned about how energy flows from a source are those who aren't *emotionally* de-pendent on it. In our house we have a woodburning stove, some candles, and a shallow well. If the power goes out, it really doesn't change our lives much. On the other hand, friends who live in all-electric, 20-story apartment buildings find the quality of their lives noticably altered when the power fails.

In essence, the desire for control, force, or pressure within any energy collection, creation, or use system is akin to desiring constant awareness and control over the very air we breathe. Consider the implications of constantly worrying about controlling your air. Energy is a similar aspect of existence. Like air and love, energy should simply be there at all times in the exact amounts needed. We feel no necessity to take deep breaths of air to last us through the night; we merely regulate its flow.

Are you emotionally dependent on energy? Here's a little test. Imagine your kitchen contains a refrigerator that only has food in it when you're hungry and then only enough to satiate your appetite and that of any guests present at that time. If you open it at any other time, it's totally empty. How do you feel about this? Does it bother you? How much pleasure and security do you derive from your full refrigerator? Do you feel comfortable asking others to dine with you? Could you ever learn to trust such a system?

Storage, Forces, and War

The wastefulness of storage is perhaps most dramatic and far-reaching in its effects on our relationships with others. Storage creates wars whether we "store" land, people, technology or resources. Realizing that inasmuch as we do to the least, storage must also be the cause of wars between individuals. The Tenth Commandment warns of coveting; however, it's the storage or possession that produces the jealousy which may then lead to theft or destruction. The person who believes he or she has nothing to steal or hide is rarely robbed or disclosed. We store things—whether energy, love or money—because we believe the resultant force created is evidence of our power. We want to be able to control how much light and heat is present, we want to love or be loved when *we're* in the mood, we want money to enter and leave our checking accounts at our command. In other words, we store to create force to give us power over other people and things.

However, we all invariably learn it's impossible to exert pressure or force upon another; we can only create it upon/within ourselves. Thus although we can say it's the force of wind or water that moves the ship, the wind and water, too, must experience change. The less resistant the object, the less it's influenced by the force. Those who dole out love or favors seeking to bind others to them can only maintain their feeling of power if the other *resists*. If the other accepts or ignores the behavior, and allows it to flow through, the "power lover" or giver finds him or herself more changed than the object he or she had hoped to overpower.

So often the concept of zero resistance—no storage—is interpreted to mean no anger, no enthusiasm, no spirit. Such isn't the case at all. When we experience zero resistance, we're free to devote all our energies to whatever we choose. As soon as we rail *against* someone, the majority of our energy is neutralized by *their* force. Think of the many stories about people doing the impossible—an uneducated young man who invents a new machine, a grandmother who lifts an automobile off her trapped grandchild. What enables these individuals to excel is that they're *willing* to give everything they have to their purpose. They're doing it because they want to— not in response to pressure from others.

When someone suggests we do something ("Sit up straight"; "Take a course in biology") are they necessarily using force? As children we all experience parental prods to "do something" which we initially resist because we don't want anyone telling us what to do. However, if the pressure or force is aligned with our subconscious desires and we're self-confident, we give in. This creates a rather intriguing philosophical question: Is a force in the *same* direction as the existing flow really a force? Or is it just more flow? If we view another's suggestion as a force attempting to alter our flow, we'll invariably resist it. However, if we can remain open and unresistant long enough to determine whether what they're suggesting is what we want for ourselves, we can save ourselves a lot of stress.

For example, as part of normal canine behavior, a group of dogs will establish a social order. Two dogs confront each other; one makes a dominate display—putting its forepaws on the shoulders of the other, for example. If the second dog goes down, rolls over and exposes its abdomen and urinates submissively, there's no fight. If the dog resists, the two will fight until one is forced into the submissive position. Is the dog that goes belly-up displaying cowardice? Indeed not; in essence it's saying "Go ahead and lead if you want to. I have other things I'd rather be doing." By choosing to display submission in this area, the dog is now *free* to experience things more in keeping with its skills, abilities and desires. For some reason, we often think every boy and girl should *want* to grow up to be President, should want to be powerful—but that's not the case at all.

Co-created Force and Personal Freedom

Unfortunately many of us erroneously equate force or power with freedom. When this occurs, the following belief structure usually results:

1. I must resist all force; force is bad and violates my freedom.
2. Any positive results that occur from what I perceive as force can't be enjoyable or beneficial in any way because they violate my freedom.

The first fallacy inherent in these beliefs is the concurrent necessary belief that others can force us to do anything. That is, we deny that any force we experience is a co-created event dependent on the full cooperation of all participants. From this we can see that by defining force or pressure as an outside influence beyond our control, all forces are perceived as violating our freedom. If the forces *reinforce* our natural flow, they're seen as indicative of our own weakness and inability to do the job alone. If the forces run counter to our flow, they're equally negative because they impede our progress. Therefore as long as we perceive force as an outside event, we have no choice but to view *all* force as a negative entity.

However, if we view the existence of any force between two entities as a co-created event, dissipating that force becomes much simpler. For example, let's think of force in terms of the pressure arising in personal relationships. We know we can always diffuse co-created pressure quite readily by decreasing resistance. Even if the other person chooses to maintain the force within him or her, the *total* force is decreased by the amount of force we withdraw or permit to flow through us without resistance. Consider the equation of two forces:

$$F_1 = F_2$$

If they're equal and opposite there's no movement, no change.

Consider, however, another form of the equation such that

$$F_1 - F_2 = F_2 - F_2 \qquad \text{(subtracting } F_2 \text{ from each side)}$$

and

$$F_1 - F_2 = 0 \qquad \text{(since } F_2 - F_2 = 0)$$

Once we withdraw a force equal to the other's projection of it, the result (resistance) is zero. For example, if we have two teams of equal strength participating in a tug-of-war, both teams experience maximum force/tension/strain, but no change occurs. On the other hand, if one team were suddenly to put their force in the *same* direction as their opponents, there would be flow and no resistance.

The only way to decrease force is to decrease resistance. When this happens, flow automatically increases, and accomplishes two things:

- It enables the other's residual pressure to flow through with minimal, if any, effects.
- It permits the previously blocked flow to manifest outwardly.

Insomuch as force, savings, stored energy or restricted love exist, there must be a barrier. If there's a barrier, then there's no flow. If there's no flow, there's no energy, no love, no life, no purpose, no *raison d'être*. It's that simple.

As in the previous chapter, let's end this one with a hypothesis: Suppose every nation chose to make its resources—minerals, oil, gas, money, talent—available to anyone, but only to fulfill the other's current needs. Would this reduce world tensions? We suggest it would. Would it eliminate war entirely? No; because there will always be people who want to fight, if only to experience combat and learn that it has no purpose.

Our true and everlasting peace, like our freedom, can only be accomplished on the basis of individual choice. Those who want to live free of war are able to do so just as those who want to fight may choose to fight. That's the nature of continuous flow. It's only when storage and its resultant force enter the picture that we come to believe that all must fight, submit, or do this or that to assuage the beliefs of an uptight, pressurized, forceful few.

11 | The Eighth Principle

The Eighth Principle: Mass is both created and destroyed.

Before we can understand this principle, we must distinguish between matter and mass as we define the terms in rotational physics. Actually, our definitions don't deviate much from the way current physicists think of the two. Let's begin by considering our full, multidimensional $_7\alpha$ spindle:

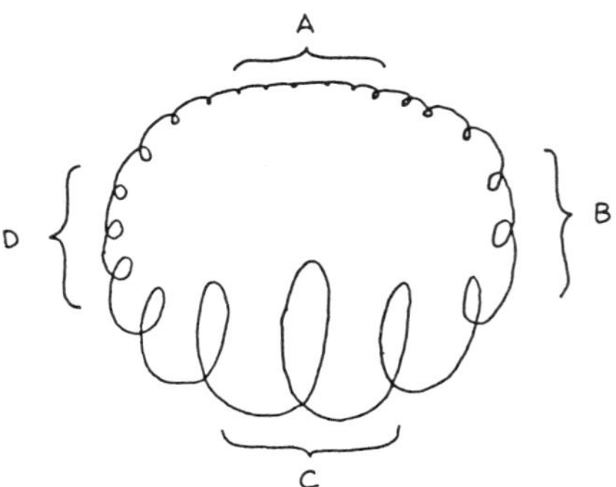

In our diagram, the area A represents the $_7\alpha$ in a configuration that is more spin than linear velocity; C represents more linear velocity less spin, and B and D, the intermediate forms. In rotational physics, the entire spindle, or ABCD, is defined as matter. Within the spindle, then, we have the continuum:

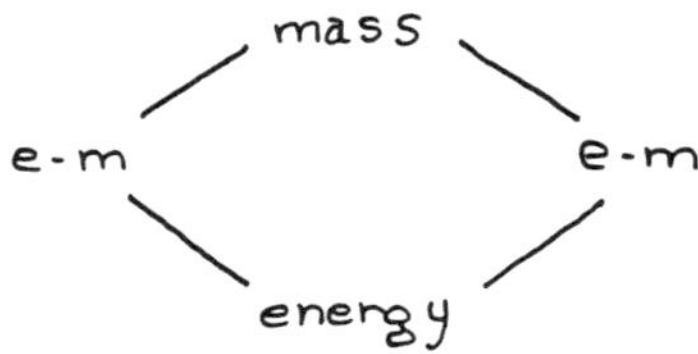

Are we saying that spin energy is mass? Yes and no. Spin energy is spin energy, but it has the *potential* to be *perceived* as mass. Think of a very long piece of very fine thread; if we wad it up in a loose ball, it's easier to see. If we then wad it up into an infinitely compact ball, it becomes harder to see once again.

There's no such thing as absolute mass. Physicists do speak of a "rest mass", noting that an object increases in mass when moving; this increased mass isn't usually measured in terms of weight or density but rather in electron-volts.* We say not usually, but it can be. For instance, an electron-volt (eV) itself can be equated to energy,

$$1 \text{ eV} = 1.52 \times 10^{-22} \text{ BTU} = 4.45 \times 10^{-26} \text{ kilowatt hours}$$

or to mass:

$$1 \text{ eV} = 1.78 \times 10^{-33} \text{ gram} = 1.97 \times 10^{-39} \text{ ton}$$

Just looking at the units—BTU, kilowatt hours, gram, ton—we see mass described in terms we associate with heat, electricity or light, a loaf of bread, or coal.

What is Mass?

So the question is: What makes mass "mass"? Looking at our spindle,

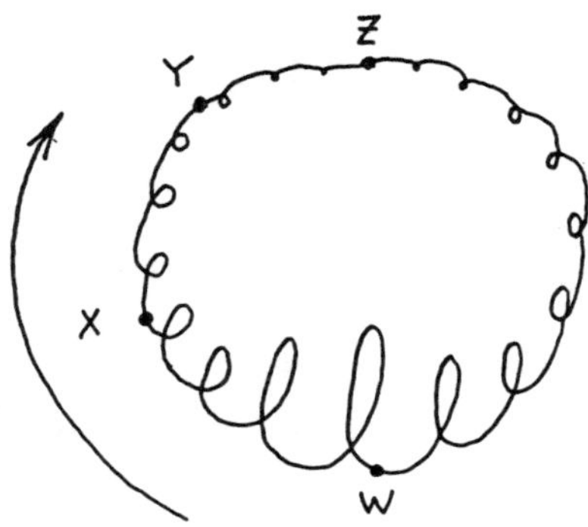

---

*One electron-volt, a unit of *energy*, represents the energy gained by a particle having one unit of charge moving through a potential *difference* of one volt.

point W is obviously maximum linear velocity. As we move clockwise, where does mass appear? At X? Y? Z? The answer lies with each individual because mass is perceptual and each person's level of perception is different. To be sure, anyone—a physicist, a chemist, a police officer—could make an arbitrary definition that mass "occurs" at a point halfway between Y and Z. This may be a handy reference for abstract calculations, but it doesn't help each of us individually.

Mass, then, must be sensed—seen, felt, heard, smelled, or tasted—for it to be real, and the point of recognition of each sensation differs with each person. The reason a sighted person can't fully describe the color blue to a person blind from birth is because blue has no reality or mass to someone who can't see. How many times have you searched unsuccessfully for something, only to have a friend point out that the pen you're looking for is right on the table in front of you? Because you weren't attuned to "pen" before it was shown to you, it wasn't real. You could have gone into a court of law and sworn under oath with 10 lie detectors connected to you and a Bible under each hand that the pen wasn't there. And *for you* it wasn't.

Try this experiment with a dozen or so friends. Find a radio station that plays background music. Turn the volume off. Blindfold your participants and begin increasing the volume of the music. As you increase the volume, ask each person to raise his or her hand when each first detects sound. Do it a few times to establish a pattern. What you'll find is that a few people have acute hearing, a large percentage fall into an average range, and a few are "hard of hearing". The question is: When did mass or real sound begin? As our friend, Einstein, told us—it's all relative.

Mass and the Self-created Reality

Many people and a few philosophies say we create our own reality. We are lord of all we survey because we're the ones who make it real. Look around you right now: That's your reality. What you perceive makes it real. If you can't perceive it, it isn't real to you.

Does this mean that each person creates his or her reality uniquely? Yes. Then does that mean that all is chaos? No. As we've noted on several occasions, the $_7\alpha$ wants to be stable and at one with all that is. Chaos, like chain reactions, isn't natural. Nature is orderly; people intuitively want to be orderly. For example, if 100 cars and drivers were suddenly put together in the center of a big arena with no traffic rules or regulations to govern their activities, what do you think would happen? Do you really think the drivers would immediately drive helter-skelter as fast as possible, deliberately ramming each other? Of course not. True, a certain small percentage who can't handle freedom might react frantically; and another small percentage who are equally freedom-fearful may choose not to move at all. However, the majority would soon establish a predictable flow of traffic that would encompass the full range of driving abilities of the group safely, because that is the most common self-created reality perceived.

Can you make things disappear for you? Most certainly. Remember the "lost" pen that was right on your desk all the time? Most of us create and dissipate mass subconsciously all the time; however, we can also train ourselves to do this consciously if we wish. We all recognize how familiarity can enable us to overlook certain qualities in ourselves and others making them disappear for us. However, few of us realize we can always choose to dissipate those qualities we consider negative at any time. Think of what happens when you find yourself in a restaurant you find distasteful for one reason or another—it's poorly ventilated and uncomfortable, the service is slow, the food lukewarm and unimaginative—and you consciously decide to enjoy yourself. At first it's difficult to ignore the physical surroundings, but as you concentrate on enhancing your awareness of the more positive mass such as your charming dinner companion, for example, the room and the food literally disappear.

If we can make things appear and disappear for ourselves, can we do it for others? Yes, as long as we have their cooperation. Such phenomena are common in hypnotism and faith healing. The well known story of *The Emperor's New Clothes* is another example of co-created alteration of perception. And surely we've all encountered the beauty in the eyes of the beholder and love so strong it co-creates charm, intelligence, wit and physical beauty in participants others may find quite boring and ordinary.

Then is everything we perceive an illusion? An illusion, yes—a delusion, no. Most of us can't perceive the total form of something. This is nothing new; Eastern mystics have long told us our descriptions of "reality" are incorrect. We tend to believe all mass is made up of the classic "particles"—neutrons, protons, electrons, mesons, neutrinos—although particle physicists add that these tiny entities aren't solid little marbles so much as electro-magnetic probability fields. Given the probable nature of the smaller building blocks of mass, it seems reasonable that any definition of real must include all these probabilities. In other words, how you see an apple and how your eighty-year-old neighbor and eight-month-old nephew see it are all part of the apple's real form. Each one has a definite self-created idea of the apple's form, but there are most likely some characteristics about which all agree.

Therefore, we each create our reality individually but there are majority agreements without which there must be mass chaos and insanity. If you invite me to your home for dinner and serve a beautiful roast, but I see a two-foot long pipe wrench on a serving platter, we're going to run into some difficulties. You would eventually call me nuts and I would quite possibly have equally strong words for you.

Have we no way to predict *how* or *when* perception will occur? None at all. In *Primer* and in chapter one we mentioned that 80% of the sighted population registers "light" when they perceive 10^{13} $_7\alpha$ photons of nominal size (10^{-13} mm) per mm^3. We can then use our knowledge of statistics to analyze how an entire group will react. We can't forget the rotational variation of Heisenberg's Uncertainty Principle however: If we know everything about a group, we can know nothing about a single individual and vice versa. If we show one hundred people our apple and ask them to select a colored square that matches the apple's color, we'll get a predictable range of responses. However, this in no way helps us predict what color any single individual will choose.

Mass and Perceptual Choice

Based on this, we can say mass is both created and destroyed because it's within each of us to accept or deny our perception of it. The "thing" itself can't be created or destroyed because it always has an energy component somewhere, whether we choose to consciously acknowledge it or not.

Imagine a rock falling from A to B:

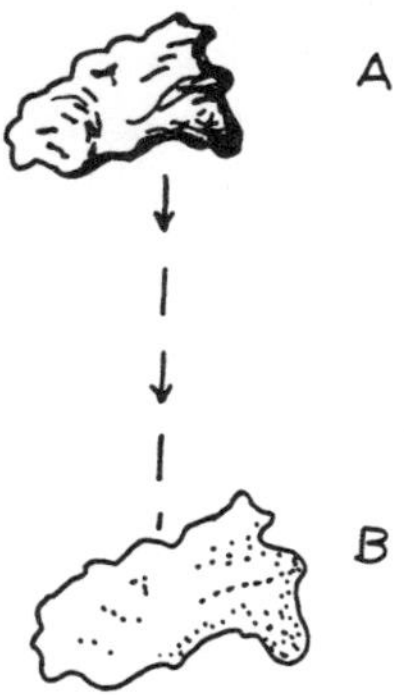

Viewed linearly we say the only changes that can occur in the rock are governed by influences crossing its downward path. If a bird or strong gust of wind crosses its path, they can alter the rock's course and structure. If they occur in paths parallel to AB, their effects are inconsequential. Thus, viewed linearly the opportunity for change is limited to an infinite number of points on a line between A and B.

If we view our rock as a rotational part of an infinite number of other both larger and smaller rotations, its opportunities for change are quite different. Recognizing our rock has a concentrated spiral configuration which we perceive as mass and also energy components, both greater spin as well as greater linear velocity,

stretching infinitely in all directions, its ability to influence and be influenced by others, indeed all others, must be infinite. The shift from linear to rotational perception enables us to see how mass is created and how some mass appears more lasting and solid.

The human race has long been fascinated with both creation and destruction. We wonder if the universe *was created* and, if so, whether or not it will *be destroyed*. Nearly every science and religion believes the universe was created by a someone or a something. Religions tend towards a creative god or gods; science looks at creative probabilistic, mechanistic events like the Big Bang. Suppose both are right and both are wrong at the same time. Let's postulate some conditions, apply rotational theory and see what happens.

If we each create our individual realities, then to some degree we must all create collective realities. The collective reality that most of us are now aware of is this universe. Hence, we all created or, more correctly, co-created it. Because we all possess an intuitive knowledge of and relationship to all that is—our God Within or omniscient and omnipresent state—this universe is indeed created by god(s). And what is the "physical" nature of this beginning? We're given a clue in the Biblical command, "Let there be light!" We know light begins with the combination of two $_7\alpha$ creating a pair. Many scientists believe the universe began with a lump of very small, infinitely dense material. The $_7\alpha$ shrunk to its smallest size or greatest spin could be thought of as being infinitely small and infinitely dense. These scientists also believe this heavy, little ball then expanded very rapidly creating lots of light and heat—more

$_7\alpha$. We need only expand our high spin $_7\alpha$ to create these other forms, and we know it takes only one entity to create a streaming effect.

Where do the additional $_7\alpha$ necessary to "stream" come from? Here the scientists stop and the theologians begin, each seeing the other's area of expertise as alien and unapplicable to their own. The $_7\alpha$ that create this or any other reality come from other realities— from heavens, hells, Nirvanas, pasts and futures—from forms that are energy-like to our mass forms here. Is it any wonder, given our intuitive awareness of our infinite nature, that the concepts of both creation and destruction or death are so alien and terrifying?

Until we fully understand and accept our role in creation and destruction of ourselves individually and our entire reality as a whole, we'll be plagued by doubts and limited in development. When we put the creation and destruction of mass somewhere outside or beyond ourselves, we divorce ourselves from both our own creation and destruction and that of all around us.

Imagine watching someone tear up a thick stack of hand-written sheets of paper; now imagine those sheets represent your life's work, the record of all you've learned and experienced here and anywhere else. How do you feel about the destruction of what you consider your personal creation? This is why the destruction of *anything* is highly dependent on the evokation of strong outside forces—patriotism, esprit, God—to place as much distance between each individual and the awareness that he or she is destroying their own creation. The unknown enemy killed, the trees defoliated, the water polluted in some far away country with an unpronounceable name are as intimately a part of our individual creations as our next door neighbor, the oaks and birches in the back yard, the pond we go swimming in, our children and the sound of our own laughter.

However, in spite of our unwillingness to recognize our individual choice to create or destroy we *function* as though creation and destruction were a choice, even if we fail to consciously recognize them as such. Let's look at a few examples. The most obvious one is war. If we didn't believe we could create or destroy mass at will, there would be no wars involving mass destruction. What's the use of destroying something if you don't believe you can replace it with something better? Where does the idea that such is possible arise from if not from one's innate awareness that such is so?

As our weapons have become more and more capable of destroying forms *beyond* mass as we recognize it, we're more reluctant to engage in war. Does this mean that the creation of nuclear weapons deters war? Not in the least. Those who seek to destroy and be destroyed simply use those weapons that conform with their awareness of their mass creative ability. Blasting some vital organ or becoming ashes or dust via various familiar routes of destruction is one thing: We all intuitively believe, or at least hope, the resultant bits of ash or dust will be sufficiently large to preserve our "code" for future regeneration. However, weapons which intimately invade and destroy the innermost core of our being don't offer us that security. What if it takes two $_7\alpha$ for each of us to be real—and a well-placed hydrogen bomb shatters us into an infinite number of *singular* $_7\alpha$? Sure, we don't cease to exist, but we've completely lost our sense of self here or anywhere else. Not a very comforting thought.

Those who seek to create life in a test tube erroneously seek to break the nonexistent creative barrier much as others seek to break the destructive one with nuclear weapons. We've mentioned before how the extremes tend to give meaning to each other. It's interesting to note that those who are pro-defense, pro-war and destruction are often simultaneously the most fundamentalistic relative to creation. Although they advocate the expenditure of trillions of dollars to create countless different ways to destroy people and property, they vehemently resist abortion, genetic engineering or anything that smacks of humankind's ability to create life as readily as it seeks to destroy it.

Why such an obvious and blatant paradox? Because we're running out of excuses. As long as we refuse to accept our ability to create and destroy, we can maintain our belief that we're not creatures of free will. As long as we fail to accept we're free, we'll fight wars for freedom. To fight, or feel the need to fight, to assert freedom is the finest evidence one doesn't consider one's self free. Those who recognize their freedom to create and destroy all mass around them have no need to fight. Those who prepare the most for war, those who see constant outside threats to their freedom are imprisoned by their own beliefs. What's unfortunate, of course, is that when such fearful individuals gain positions of power, it's easier for the indifferent 80% of the population to choose to *temporarily* accept that philosophy.

Many people believe that if all had the power to destroy at will, there would be widespread devastation; three-year-olds would whimsically vaporize buildings, disgruntled Viet Nam veterans would level Washington, an angry driver would disintegrate the slow moving truck holding up traffic. People who express this fear are only acknowledging half the picture. To have the power to destroy at will has no meaning unless one can create at will also. For every three-year-old who desires to vaporize a particular building, another chooses to maintain its presence. The reason the concept of nuclear defense is so impotent is because there's no creative force to balance it. What's the use of destroying something if you don't believe you'll be able to replace it with something better? Can anything perceived as real be so bad replacing it with an empty void is preferable? Such thinking is reminiscent of the petulant child who sends his playmates home because they refuse to see or do things his way. Sure he won, but what did he win?—loneliness, incredible loneliness.

Co-created Creations and Destructions

When viewed rotationally we can easily see why all creations and destructions must be co-created events, even though each individual is capable of accomplishing them him or herself. It's the presence of the second individual that gives validity or meaning to our creation or destruction. So strong is this awareness that few complex species are even capable of reproducing their own kind singularly. Once the organism becomes sufficiently intricate and its choices magnify accordingly, a built-in system of validation arises. In most species the birth of an offspring denotes a "creation" resulting from the interaction of participants. However, the need to tag or label one's offspring for life is a peculiarly human trait, again arising from our unwillingness to accept creative function as a matter of individual choice. We're willing to accept that Harry and Helen Horvath choose to conceive little Harry, Jr; however few of us are willing to recognize that Harry, Jr. is the primary source of choice in his own creation.

What we often overlook is that the birth of a child is in absolutely no way representative of the race's ability to create mass. Children are unique composite beings of energy (spirit), e-m (mind), *and* mass (body) and as such can be created or destroyed by no one or nothing but their own choice. The conscious or subconscious choice of a parent to participate in the entry of an infinite entity in mass form is merely to agree to serve as a vehicle, a train choosing to stop at a particular station to pick up a particular passenger, and then discharge it at a mutually agreeable time and place. The fact that one train picks up the physiological form does not mean the spirit and mind of the infant can't "come through" another. That's why some children are the "spit and image" of their parents whereas others have only the faintest similarities.

Similarly it's possible for a child to choose to experience only part of the trip, disembarking during the first, second or third trimester of pregnancy or at nine, or fifteen years of age. Parents or others who feel victimized by such events fail to accept the co-created nature of this experience. They often acknowledge it to the point of recognizing their role as abnormal, that they did something wrong, and choose to live with crushing guilt. However, that's an extremely limited and distorted view. Parents and children, like $_7\alpha$, work in harmony to co-create things they both wish to experience. Compare the response of parents who use the death of a child to project them into greater awareness of their love for themselves and others versus those who use a similar event to prove themselves victims of a rotten world. Which ones are free? Which ones have taken a seemingly negative event and used it for the benefit of all mankind? Which child provides the greater gift?

If our children are proof of their creative ability and not ours, where can we turn for proof of our individual creativity? The arts— painting, sculpture, music, literature—are our most readily available proof of our ability to create mass by choice. The artist, musician, dancer literally begins with nothing, no mass—merely a feeling, pure energy—and creates mass. We can't say the blobs of paint that became the "Mona Lisa" *are* the "Mona Lisa" any more than the sacks of concrete are the bridge. They're merely paint and concrete. When the artist or builder conceives the painting or bridge, the mass is built in the mind first from the pure energy of creation, then given form. This natural and necessary progression of

$$\text{energy} \longrightarrow \text{e-m} \longrightarrow \text{mass}$$

$$\text{inspiration} \longrightarrow \text{mental image} \longrightarrow \text{real object}$$

is vital to all creative processes and the reason why we must never stop dreaming. Without dreams, without the energy and e-m, there can be no mass. And ours is a *mass* reality.

Imagine yourself planning to build something—a saw horse, a patio, a rock wall. See the object in your mind and follow the creation through to completion. Was it necessary to surround yourself with images of all the raw materials and tools in order to plan the construction? Isn't it possible that having those items could actually limit your creation? For example, suppose you're a do-it-yourself klutz like us, and imagine a brick patio with railroad-tie terraces, flower gardens, a small waterfall tumbling over rocks in one corner, and furnishings made of rough-hewn beams. Suppose in order to imagine this you had to imagine all the necessary tools and steps in the creative process. Like us, your knowledge of tools is so limited you wind up having to build your patio with a hand saw, hammer and screwdriver. How does having to use tools and methods from your own conscious *mass* experience affect the quality of your dream patio? It's quite limiting, isn't it? You wind up building something to suit what you have or know rather than creating what you want.

Such is the nature of those who point to raw materials such as rock, sand, or wood as proof that mass can't be made or destroyed except by gods, even though we live in a world where nuclear energy has already nullified the destructive end of that belief. By saying the bridge is merely another form of concrete, and concrete merely rock and water, they're able to follow the "raw" material back to its point of creation. Then they must decide who brought that rock from nothing to its current mass form. God? Chance? Me?

At this point many choose not to know at all. We select some arbitrary point of creation and hope enough others will agree with it to make it real. Sometimes it works; more often it doesn't. More often someone comes along and wants to push our arbitrary point further back or forward. Any time anyone says we ought to change it, we *can* change it, they simply point out what we were trying to avoid. Our awareness of where creation, real life, or mass begins is up to each one of us.

So in a mass reality the inhabitants create and destroy mass on a regular basis. Unfortunately in this reality the majority belief says that we must destroy (fission) or create (fusion) mass in order to create energy. As we'll see when we discuss the Tenth Principle, energy *can't* be created or destroyed. That being the case, if we want energy we must funnel it, channel it, concentrate it, collect it. We need do nothing more because energy simply *is*.

In order to get from our awareness of ourselves as the creators and destroyers of mass to our role as collectors and filterers of energy, we need to take a rotational look at reactions. In the next chapter we'll see what happens whenever any reaction occurs.

2 | The Ninth Principle

The Ninth Principle: For every (re)action occurring in nature, some change in either mass or energy occurs.

This principle may seem like a simple restatement of a well-known fact, but it's surprising how many people view (re)actions in terms of change in *position* rather than *state*. If a block at A moves to B,

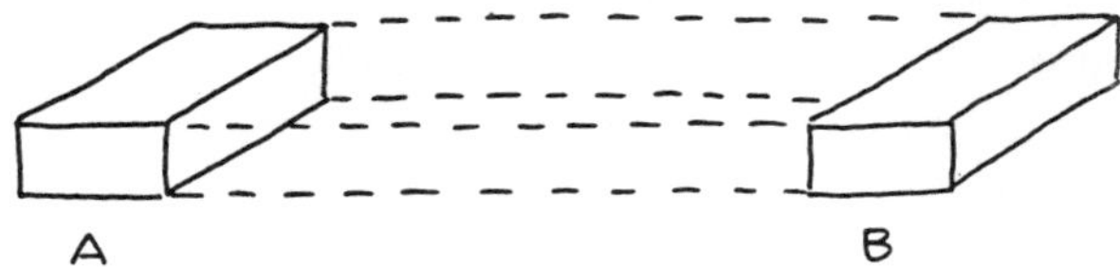

many of us say the block remains the same; that is, there's been no change in block A, or Block A = Block B. However, *any* change, including a change in position, necessitates some change in the energy and/or mass state of the object. Therefore, regardless how perceptually similar A and B appear, A cannot equal B (A $\neq$ B); the best is an approximation, A $\approx$ B.

Unfortunately many scientists and engineers tend to overlook or misinterpret the change inherent in every (re)action occurring in nature and our universe. Remember Newtonian mechanics? For every action there's an equal and opposite reaction. That means that if we push down on a table with a force of 10 kilograms, the table resists or pushes back with an equivalent force. This implies a force balance

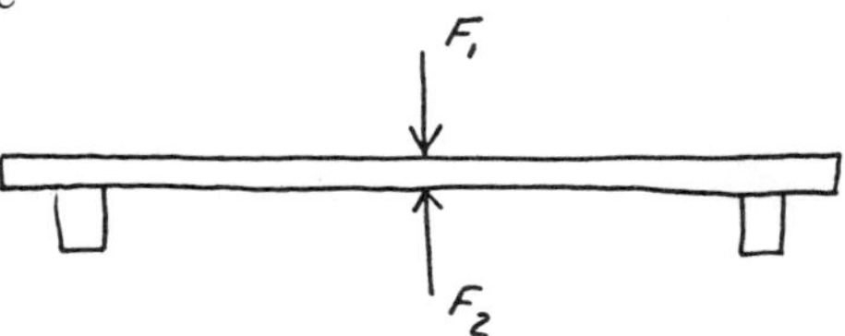

with little or no consideration of mass at all—unless the table collapses. When applied to human philosophy, this becomes "an eye for an eye".

However, a reaction by definition is the *opposite* of an action. In terms of the natural rotational sequence, if we remove energy or motion we create mass and vice versa. Therefore, if we remove or take anything somewhere, someone or something must be willing to replace it. Also, if we accept something, someone else must be willing to give that something up. If we tug on one corner, the whole universe moves.

No action or reaction occurs without change. Even if the overall system remains stable and produces no *perceivable* change, there's change nonetheless. If we have a collection of 10-kilogram weights and replace one weight with an identical counterpart, the total number, the total weight, and even relative position of the weights in the group remains unchanged, but the weight removed isn't the same as the one now present, nor is the resultant collection the same.

Engineers expect every atom and molecule in a steel beam to remain in place and function as though the forces acting on an object are unchanging. Even though the forces result from the constant arrangement and rearrangement of infinite numbers of $_7\alpha$, engineers see them as so many fixed "weights", aligned and never (or rarely) changing.

While such rigid thinking may have advantages in a linear-based society, it also limits creativity. As long as the electrical engineer believes all electrons are the same and can be trapped in a conductor, and the chemist that the oxygen in her molecule of H_2O is always the same oxygen, both relinquish multidimensionality *and* probable alternate manifestations in favor of absolutes. In such a way our chemist not only limits her creativity—"What can I do with this molecule of water?" rather than "What can I do with the infinite manifestations of two infinitely variable hydrogens and one infinitely variable oxygen?"—but also her knowledge.

To believe the movement of a block from A to B is the result of external force is as erroneous as believing the birth or existence of a child results from its passage through the birth canal, or the automobile comes into existence by virtue of its trip from the manufacturer to the showroom. On the other hand, to believe the child or vehicle which embarked on a journey is identical to the one arriving at its destination is equally erroneous. However, this is the logical conclusion if we see motion as a result of external forces rather than internal change. Nothing moves that doesn't want to move, just as nothing stops that doesn't want to stop. The action/reaction of nature and all things is *within* each system. Only in an individual willing to accept change, to convert mass to energy or vice versa, is either motion or stability possible. Remember: Unless there's change, no motion or stable state can be perceived.

Self-originating Change

Accepting there's some change in mass or energy for every change occurring in nature presents problems for most in terms of accepting those changes as being *self-originating*. Few question the "forces" of nature on the environment creating erosion, crumbling, oxidation. However, few can easily accept those forces as originating *within* the soil, mountain, or metal as a function of choice.

When engineers or architects design buildings, the nature and magnitude of outside forces are their primary concern. It's assumed the weight of the building and its contents create forces based on the gravitational attraction of the earth. Beams of sufficient size and number to counter these forces are then designed to support the structure. Similarly, other external forces are reckoned with—wind, rain, snow, ground tremors. Safety factors are applied which in essence make our building capable of holding back these external forces many times over, the idea being these buildings should last forever, or at least a very, very long time. Suppose rather than applying the concept of linear force, designers made use of the concepts of rotational engineering focusing on the *internal* relationships (V_S/V_L) of the components. The changes in mass and energy would be predictable and therefore

- The structure would be lighter and more economical to build because it wouldn't need safety factors.
- There would be no buildup of forces, like the weight seen as constantly pushing down on the lower members. Any outside forces would literally pass through the building.
- The building wouldn't be rigidly constructed with beams riveted to other beams, but would have a degree of flexibility built into it. This would eliminate failures from fatigue.
- It would last a long time indeed.

To see how this works, let's create a mile-high building in the shape of a pyramid and examine it in terms of changes in mass and energy. The traditional linear view of the structure prevailing against the ravages of nature is

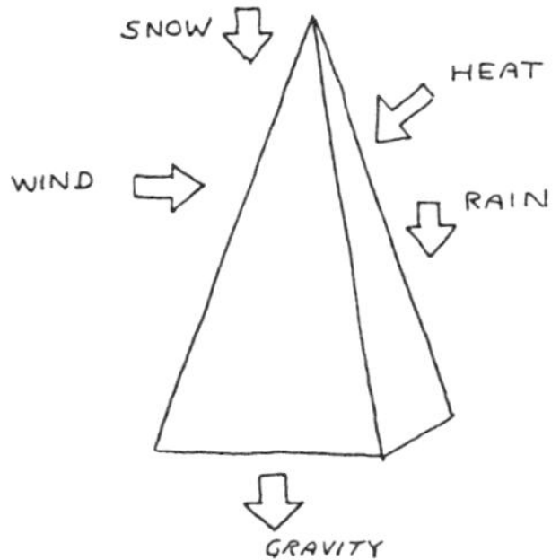

Rotationally our building looks like this:

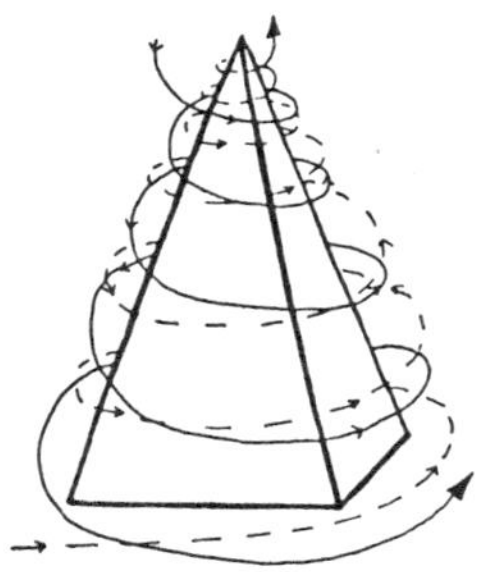

The external forces create a double helix which, because of the great flexibility built into the structure, pass through it rather than around it.

Although such a view offers more stability and greater range for creativity than the more linear one, it doesn't attribute any force-creating potential to the structure itself. Obviously if such is the nature of the external or applied forces on the structure, in order to be balanced or at one it must generate its own equal but opposite forces—which indeed it does:

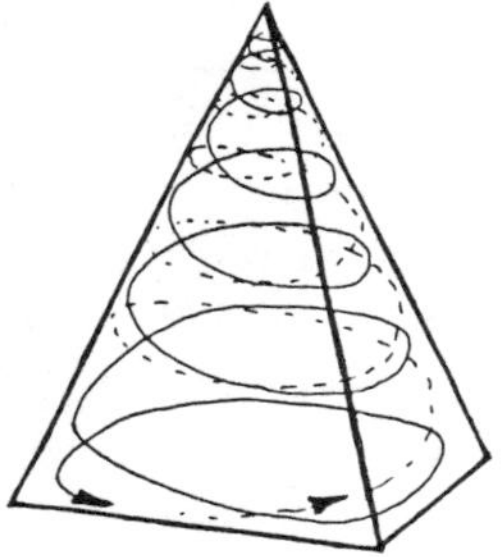

Therefore, we can view the system as two sets of complimentary rather than antagonistic forces working together:

- An external component acting on the structure.
- An internal component acting on the environment.

If these two components are equal, there's no net change and the structure is totally stable: When two equal and opposite forces are present, there's no change.

The human parallel of this phenomenon is most obvious in those having strongly diverse opinions. Although they may spend a great deal of time in their push-pull contest, neither gives an inch relative to the other. Each expends a tremendous amount of *internal* energy but creates no external change, no change in belief, no motion. The one who chooses to stop the argument breaks this cycle. Therefore, we can see how the one who walks away from an argument manifests more energy than the one who remains fixed.

Often the one remaining accuses the other of cowardice because of his or her awareness of this greater power; that's merely the mirror phenomenon—the projection of their own beliefs and fears about themselves onto the other.

Internal and external forces co-create change and produce changes which must, by definition, affect all that is. This is a phenomenon often overlooked by more linear, logical thinkers. We often view certain changes within ourselves as being forced upon us because others won't change—"I'd prefer to make a change in my life, but society won't change its rules so I must stay the way I am." "If I had my choice I'd _______; however. . . ."—the implication being, we're being forced to do something against our will. Insomuch as an individual refuses to accept the choice to change, he or she limits the ability of society to change. Because all is connected, to believe you can hold fast and wait for others to latch on to your beliefs is folly. That amounts to all the rigidity and eventual decay inherent in structures designed to *withstand* linear, outside forces rather than exist in peace with the natural rotational forces within and around them.

Rotationally, then, the *only* way to change any part of society is to change the self *first;* as soon as the single $_7\alpha$ changes, the entire universe responds. If the $_7\alpha$ doesn't change, it's agreeing with everything as it stands. If it doesn't choose to change from within, then its only other option is to change *in response to* other $_7\alpha$. Think of a person you know who's resistant to change; we'll call her Linda. Because Linda chooses not to change internally, she sees her life as a constant progression of responses to others, often negative responses because she sees others as trying to force her to do things—"Linda, go to lunch with us." "Dear, would you iron my shirt?" "Mommy, make me a sandwich." Because she refuses to acknowledge that she must initiate an internal change to stack the decks in her favor, Linda is little more than a participant in other people's probabilities. If Linda recognizes she goes to lunch, irons shirts, makes sandwiches or doesn't because that's what *she* wants to do, how or what others ask of her has no power/force over her at all. If she wants to iron her husband's shirt because she likes him to look neat and well-groomed, how he asks her to do it, whether he asks at all or even notices her perfect ironing job is immaterial. Such considerations are only important if she sees the task as her spouse's means of exerting force on her rather than a choice she makes herself.

Think of an engineer viewing the winds surrounding our mile-high building. How does his view differ if he sees the building under wind attack versus seeing it as an equal but opposite force which can work with these other wind forces? In the former case we're involved in an impotent and nonproductive push-pull contest. In the latter, the engineer alters the internal forces to align them with the external. In such a way not only does he maintain the stability of the structure, he also creates a system whereby the combined energies may be collected and utilized.

So Linda looks at her impeccably dressed mate and takes great pleasure in her contribution to his appearance. Knowing she did a good job by choice for herself makes her feel good, stable, at one. If her spouse shares her enthusiasm for her work, fine; if not, it in no way lessens the value the work has for her.

Force Imbalance and Energy Production

Regardless of their magnitude, if two equal and opposite forces act on one another, there's no energy created, no change, no motion. In order for energy production to be initiated, there must be at least a momentary imbalance in force. We may accomplish this by lessening one force or increasing another, although the former is invariably more efficient and less time consuming. Once the change initiates the flow, it is perpetuated via the streaming effect.

Changes in mass and energy occur constantly in nature. Dead plants and animals decompose and produce heat; energy from the sun is taken by plants for their growth. If we want things to remain stable, we must allow the changes to equilibrate, converting equal amounts of energy to mass as mass to energy. Unfortunately, our current energy focus is on the conversion of a great deal of mass—

coal, oil or nuclei—to produce a pitiful amount of energy. Furthermore a secondary mass form, waste, is created which is of no use to us at all—either as mass or a future energy source.

The resources of the earth are far more than simply something for us to waste at our pleasure, something for us to force to change unnaturally to fulfill our desires. If we continue the way we're going, the major problem won't be the depletion of oil reserves, coal, or natural gas; that would be relatively minor in and of itself. The result of needlessly converting large quantities of the earth's resources to energy is that we may permanently alter the balance of the planet, turning a beautiful world into a wasteland. It would be a tremendous and unnecessary tragedy to convert our limited mass to energy when energy is present in its natural form for the taking. All we need be willing to do is change.

How do we know the energy's there for the taking? Because, as we'll discover in the next chapter, energy is neither created nor destroyed.

.3 | The Tenth Principle

The Tenth Principle: Energy is neither created nor destroyed.

To say that energy is neither created nor destroyed has about the same impact as saying the earth is round; it's a convention accepted early in our education with little or no thought given to its meaning. When we couple this principle with the Eight Principle (Mass is both created and destroyed.), the logical question is: Why can mass be created and destroyed but not energy? Again, the answer lies in a clear definition of terms, terms which to date have largely been ignored. Mass is the unit or state of reality. Unless something has sufficient $_7\alpha$ concentration to trigger our perception, it's as though it doesn't exist. We are not saying it doesn't exist; we're saying it doesn't exist *to us*.

Energy, however, is a different matter. Although we have many indirect ways of measuring it, we can't feel, taste, see, smell, or hear energy *per se*, only its effects. Thus, if mass is the unit of *here*, energy is the unit of *there*. This doesn't mean that mass only exists here any more than energy only exists there; rather, it denotes the more multidimensional nature of the energy state.

Examples of this principle are numerous. Consider an object that's mass to you by virtue of its effect on two of your senses: an apple pie, fresh from the oven, which you see on the table and smell. When these senses aren't stimulated, when there are insufficient $_7\alpha$ to stimulate the receptor sites to register "pie shape" and "apple pie odor", the pie isn't real to you. A test for reality is simple: Would you bet your life, your reality, on it? If you could only see the pie, would you be willing to bet it was apple? If you could only smell it, would you be willing to bet it was an apple *pie* and not a turnover or a cake?

So regardless whether mass is present or not, energy like love always is. For example, you can imagine the taste, look, texture and smell of Mom's apple pie any time, anywhere. There are many ramifications of this in terms of probabilities which could lead to arguments and discussions similar to those having angels dancing on pin heads. All mass, while finitely bounded by time and space, has an energy content which is infinite. One of Mom's pies can only take up so much space and last so long, but the memory of it can last forever. Most importantly, the corollary is also true: All energy has a mass component somewhere bound by space and time. Our desire for the pie can lead to its creation for, and consumption during, a particular Thanksgiving Day celebration at 121 Main Street in Marion, Ohio. This must be if all is infinite and connected.

Of course, the corollary means that whatever we perceive as energy here has a mass or real form somewhere else. Think about that for a moment: How energy is collected and used here must affect those forms in other realities, just as "their" utilization of energy affects our form here. Suppose every time you turned on a light you affected the milk in a town across the state; every time you started your car, you changed the purity of their water. Would such awareness influence the form of energy you chose, how you collected it, and its ultimate use? Indeed it should, but the fact remains, it doesn't. For our energy choice, collection, and utilization does indeed affect others, but we ignore it. If pollution occurs here among those we know and see, what hope is there for countless millions of unknown realities with their unknown inhabitants which we can't see? It's been said before—"How can you claim to love God whom you cannot see but not humankind which you can?"

The terror of nuclear weaponry and energy producing plants isn't so much what it does or can do here, but there. A large-scale thermonuclear war would certainly end this reality, but that's not the issue multidimensionally, because mass realities are "born" and "die" on a regular basis. A thermonuclear war on the scale the super powers are arming for would begin an inter-reality chain reaction that would make the Big Bang seem like a whimper. *That's* the real issue and it was mirrored in Einstein's unheeded warning that the detonation of the first atom bomb might have triggered a massive, world-wide chain reaction—even if it didn't do it *here*.

The purpose of multidimensional awareness isn't to impress you with your inconsiderate nature or ignorance, but rather to point out how the first step to a proper energy philosophy and technology must originate with individual enlightenment. By implying matter (and defining it mass *and* energy) is indestructable, modern physics has conferred immortality *a fiat*. That is, rather than defining mass reality as within the creative province of each individual and energy within the province of all that is, matter is defined as some gigantic pool from which we pull our mass and energy until it runs out. The vast majority of current energy policy really has little regard for the environment; it's primarily concerned with the development of a cheaper source.

In other words, current energy policy views energy as something here which we "discover" or create and destroy at will. Although we need little knowledge to recognize the effects of the sun's energy on the earth, little thought is given to the effect of our earth-produced energy on the sun.

Energy collection should be compatible with the environment in which the collection takes place. The collector that's appropriate for northern Canada may be quite different from the one for Mexico City. We need only look at the avearge nuclear plant to see it is as alien to its environment as a skyscraper in an Iowa cornfield. Remember: Regardless where any energy producing or collecting device is located or how it works, we always get out what we put in—garbage in, garbage out.

Energy and Infinity

The idea of an infinite source of energy is certainly quite a concept for us linear thinkers who claim that the infinite, or infinity, is beyond our comprehension. So in order to be able to comprehend a principle dealing with infinite energy, we must first be able to comprehend infinity. The physicist and author George Gamow* created a fascinating infinite analogy which is extremely helpful.

Imagine you're at the reception desk of the Finite Hotel. You ask for a room and the clerk responds, "I'm sorry, all the rooms are full." You go across the street to the Infinite Hotel which, quite naturally, has no limit on its number of rooms, and make the same request. The Infinite Hotel clerk replies, "All the rooms are full, but if you'll just sign the register, the bell captain will see to your luggage." Now an *infinite* number of new guests arrive at this full but Infinite Hotel and they, too, are shown to their rooms, presumably by an infinite number of desk clerks and bell captains.

How can this be? Is it beyond comprehension? If it is, it's because we've trained ourselves to think linearly and finitely and therefore simply ignore those probabilities that represent events outside these limits. When circumstances force us to, we often refer to them as miracles; we accept such events as real but not within the normal range of probabilities. This is unfortunate because rather than looking at any exceptional event as merely a stretching and redefinition of the norm and increasing our own experience accordingly, we tend to say, "It's beyond me." We sit in our rooms in the Finite Hotel and watch the infinite progression occurring across the street, but never get up the courage to check in ourselves.

Defining Mass in Terms of Energy

When we speak of our inability to either create or destroy energy, it's primarily a matter of definition. If we consider the change of mass as it becomes more and more energy-like (spin, or V_S, being given up in favor of linear velocity, or V_L) or appears so to us, we *could* say mass is also infinite, neither created nor destroyed. However, to do so, creates unnecessary confusion. Mass is a totally arbitrary concept determined by each individual's perception; energy is beyond perception and simultaneously the source of perception itself. In fact, to the physicist it's often far simpler to

---

*In his wonderful book, *One, Two, Three. . . . Infinity.*

eliminate mass altogether from any calculations, thinking of it in terms of being "less energy" instead. Such a process merely reduces a multidimensional system to its lowest common denominator.

Imagine a system based on the colors blue, violet, and red. Rather than expressing them mathematically as three separate quantities where A = blue, B = violet, and C = red, and then measuring some unknown to discover it contains x units of A, y units of B, and z of C, doesn't it make more sense to express all three colors in terms of one if possible? Thus, if we select blue as our standard, violet can be represented by A/2 because our violet is half red, half blue, and red as A/0 ("zero" blue). Now, let's calculate the composition of our unknown using this form and rotational mathematics. Although rotational mathematics differs from linear mathematics somewhat, it's simple enough to follow.

In linear mathematics, the total units of all three colors in the unknown is represented by

$$xA + yB + zC \qquad \begin{array}{rcl} A &=& \text{blue} \\ B &=& \text{violet} \\ C &=& \text{red} \end{array}$$

Expressing this rotationally all in terms of blue we have

$$xA + y(A/2) + z(A/0) \qquad \begin{array}{rcl} A &=& \text{blue} \\ A/2 &=& \text{violet} \\ A/0 &=& \text{red} \end{array}$$

The second equation is the kind we get into when we evaluate the total mass, e-m, and energy states of anything, regardless whether our "blue" is the mass or energy state.

For simplicity's sake, let's temporarily forget about the e-m state and define an object's mass and energy states both in terms of energy. For example, we may define the energy component of a table as A and its mass component as A/0. Now let's see what happens when we assign a numerical value to A. Because the table isn't moving, nor are parts of it appearing or disappearing, we may say that whatever energy is present in the table is all there *here* and *now*. Therefore, we may set A equal to one.

How come the amount of energy isn't infinite?—Because we're defining our table in terms of the form(s) in which it exists in the finite here and now. Imagine a sealed plastic cube filled with water immersed in a tank of water. Although the water within and surrounding the cube have the same form, the walls of the cube keep them separated. Although the amount of water in the tank may be infinite, that within the cube is not; however, whatever is within the cube represents all (1.00 or 100% of) the water inherent in the cube.

When we set the value of A or the energy state of our table equal to one, the mass state then becomes 1/0 which is the same as infinite. How can the mass in our table be infinite? First of all, remember we're talking about our table as it exists here and now. That means that, unlike energy forms of the table which may exist in other realities, absolutely all of that table's mass is present right here; it has no other mass anywhere. Relative to all its other potential forms, this is its maximum mass expression which functions as relatively infinite compared to its other states.

Imagine our cube of water, only this time the walls are a different form of water—ice—rather than plastic. Although we may say the potential to become ice exists in all the water surrounding and within the cube, that which comprises the wall is all the ice that exists anywhere in this system. Compared to the zero ice state everywhere else, this is the maximum or infinite ice state.

The key to understanding how this works is to keep reminding yourself that all our observations of the table, or anything, must occur within the here and now because that's the only place where mass exists. Our memories of things we did yesterday, our plans for tomorrow, our dreams of ourselves doing certain things tonight are all energy forms of us. Our total infinite mass form is that, and only that, which we manifest this instant.

What happens when mass converts to energy? Let's imagine several mass-energy conversions: wood burning, water evaporating, the combustion of gasoline in an engine. When the fuel or mass is completely gone, in unidimensional physics we say

- It has *all* become energy.
- The energy—heat, light, pressure—lasts only as long as the fuel, and then "disappears".

However, no physicist worth his or her credentials would really say the energy ceases to exist. It's this awareness that led to the ubiquitous and ambiguous "*matter* is neither created nor destroyed."

Let's look at a block of wood which, among many other things, we may consider a potential "block" of heat and light energy. From the point of combustion through the point where nothing is left but a small amount of ash, the progression of its disappearing mass-ness looks like this:

Relative to the $_7\alpha$, we may say that those comprising the block become less and less different from those in their immediate environment. In other words, as the $_7\alpha$ lose spin and gain sufficient linearity they become indistinguishable from those in the naturally more energy-like atmosphere.

However, our conversion process is more than a loss of more mass-like $_7\alpha$ into an infinite sea of more energy-like ones where we lose sight of them and therefore the mass form. All $_7\alpha$ are infinite, but as we know the range of probability manifesting as mass in any one reality may be quite limited. What actually happens to our block during its transformation is that the $_7\alpha$ begin to align

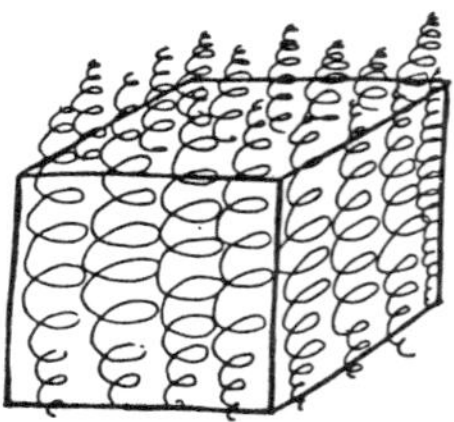

as well as stretch out and become more linear. As they stretch out, they no longer trigger our perceptual mechanisms; first we lose sight of the object, then we can no longer feel the heat. Were this alignment to occur throughout the block simultaneously, it would instantaneously disappear. However because the mass is composed of $_7\alpha$ having a variety of V_S/V_L, *when* this occurs varies. As soon as a $_7\alpha$ reaches the level of energyness of those $_7\alpha$ in the air around our block, they're drawn into it. Philosophically, this conversion from mass to energy is analogous to the concept, "He that loses his life (mass), shall find it."

Let's go back to our cube of water in the infinite sea of water. Imagine the ice walls of our cube melting; the cube, the water within, and the sea around it are now all the same. Similarly when all the potential energy in a mass is converted to energy, the mass gives way to infinite, eternal energy.

Infinite Energy Systems

We can see how the creation and destruction of mass and the noncreation and nondestruction of energy are primarily matters of definition. However, just as our definition of forces as rotational and self-originating as well as external opens new vistas in engineering, so the idea of multidimensionally infinite energy and limited mass enables us to approach familiar problems a new way.

For example, from this principle and what we've said, it's obvious that if we want to create a system in which infinite energy production manifests here, we must begin with a quantum of mass which is totally converted to energy.

Suppose instead of all the mass present in our table here and now converting to energy here and now, only half is converted.

Now our energy state A is reduced to ½ and the original 1/0 mass state becomes ½/0. In addition to cutting the amount of available energy here and now in half, we've also decreased the amount of mass from which any future conversion may occur. But isn't half an infinite amount (½/0) still an infinite amount? It's an infinite amount in that it's all the mass there is here and now; however, it is also half the amount of what we had before. Imagine yourself without either arms or legs; your body is all the body or mass you have, but it has both less mass and potential to do things than your fully limbed form.

Dealing with fractions or parts of infinite states is difficult because our perception by definition creates our *finite* limits. If one million is beyond our comprehension, the fact that it's one thousandth of a billion is immaterial; if we can't recognize 1000 watts of light because it's too bright, it doesn't make any difference that that's only half the amount present. However, if we need the total amount to maintain a conversion to a form we *can* perceive and use, it has a great deal of meaning. Remember when a teacher made you wait until the whole class finished their tasks before you could individually do something else? Even though the work of the whole class had no meaning—your job or reality was to do ten problems once, not twenty-eight times—your freedom to do something else was totally dependent on the output of the entire class. Until that greater task was accomplished, you couldn't do what you wanted. So this greater output became important, not because it represented something you wanted to imitate, but because its existence freed you to do what you wanted. Unless that total (infinite) amount was present, you couldn't function as you desired in your subinfinite state.

Similarly our emphasis shouldn't be on how *much* energy a substance gives off or how much mass converts to energy, but whether or not that conversion is *complete*. The teacher doesn't care if Johnny does his problems faster than Susie, or if Steve does fifteen to Martha's five; the only way each student can do what they want is if all twenty-eight complete the entire assignment.

In essence this simple analogy describes the streaming effect; unless the mass present is totally converted to energy, no infinite streaming can occur. Therefore, if we totally convert one milliliter of water, we can create more energy than that produced by the incomplete conversion of 4 tons of U-235. This has been known ever since Einstein proposed that $e = mc^2$.

Balancing Mass and Energy Conversions

The problem is how to put the expression $e = mc^2$ into practical use. The existence of an infinite streaming effect in energy isn't an alagorical or fantastic speculation, but merely the result of simple physics. We know the creation and destruction of mass maintains the overall balance of any reality. Obviously if all mass forms including life lasted forever, we would soon run out of room. The only option would be to maintain an ever-expanding world. Such was possible when the earth was viewed as flat, but that view concurrently fostered notions of dark perils and evil spirits at the periphery. (There's the old joke about Columbus beginning his voyage with *four* ships, but one fell over the edge.) However, it is much easier to expand a plane

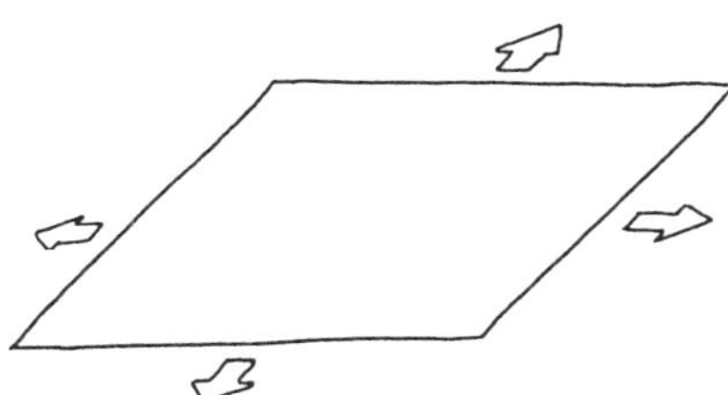

than a sphere.

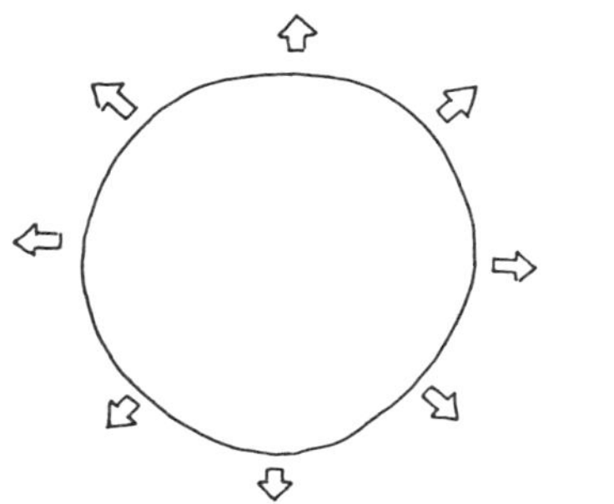

Thus once we conventionally deemed our reality round and then measured it, we simultaneously limited our creation; we became mortal. As soon as our world was no longer infinite, then all in that world lost its infinite nature also.

With the creation of a finite spherical world, some way had to be created to maintain the amount of finite mass and energy within that world. We can create such a "world" by placing light pieces of paper within the circular current of air created by a fan.

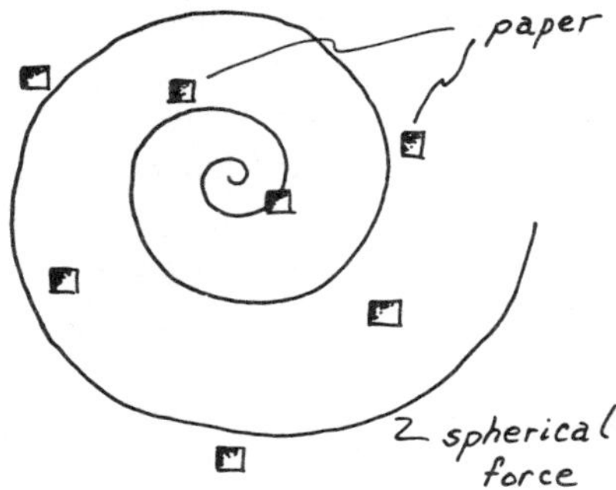

Although there are an infinite number of ways the pieces of paper may arrange themselves within our spherical field, they're limited by the available space. Eventually a point is reached where there is neither enough room nor force to hold any more paper. There are two ways we can prevent overcrowding and subsequent destruction of our spherical force-world:

- Every time one piece is put in the system, another is removed.
- Periodically some pieces leave and look for a new system.

In the first solution an optimum level is determined which is maintained on a one-to-one basis—one piece in, one out. The second solution is based on group migration. When the population is sufficiently excessive that optimum conditions give way to detrimental ones, a group of pieces leave permanently. In these ways the overall balance of the system is maintained.

If we convert our papers to mass form $_7\alpha$ and our force field to energy form $_7\alpha$, we can see how the balance is maintained. There's only room for so much mass in any reality, and there must be some system for regulating that amount.

If the amount of mass is a fixed component in a reality, how can we take a small quantum of mass, convert it totally to energy, and create an *infinite* stream of energy without destroying the balance? First of all, we must realize that the balance is maintained by mass, not energy. It's the mass that determines the reality; if we were to remove the Grand Canyon and leave a void, the entire earth would be affected. However, if we were to channel some of the infinite energy of outer space to a specific point on earth and immediately dissipate it and create mass *change*, no imbalance would result because the energy changes the *form* of mass, not the amount—imparting motion, for example.

You may recall we earlier emphasized the creation of pure energy from incoherent energy; in this chapter we add the concept of energy creation/collection from mass. Which method we use is a matter of choice. We can take sunlight, an incoherent form, and filter it to produce inherent light. We can also take water, the most prevalent mass form within the human body and the earth itself, use fusion to compress it into mass at the $_7\alpha$ level and then fission to separate it into inherent $_7\alpha$ energy forms. These are our future probabilities, and our future choices.

Now let's depart from our principles for a further discussion of probabilities, expanding our awareness deeper into the multidimensional realm.

14 | The Nature Of Probability

In chapter nine we introduced some elementary probability theory; now let's expand some of these ideas into the multidimensional realm. Everything is probabilistic: Nothing is absolute. For centuries a debate has raged around the subject of pre-determination. One camp believes that the course of human development is fixed and can't be changed. The opposite extreme, the one we prefer, holds that we're beings of choice and free will and therefore can change *anything*. The fact that probabilities even exist is proof of our free will; if we didn't have free will, we would have no use for choices. If we wear size seven shoes and there's only one pair in that size in the store, it doesn't make sense to try on all the other pairs of shoes.

When people predict the future, all they do is focus on the higher probabilities, those which are more likely to occur. Imagine two individuals observing a distant road. One not only has much better eyesight than the other, he's trained himself to look for specific clues that indicate traffic on the road. Because of this, he can see changes long before his companion. To the companion, it's as though his friend is capable of seeing beyond normal human capability. Obviously if I see someone approaching ten minutes before you do, we could say what I see in my present is actually in your future. Although it's perfectly normal within my range of perception, it seems extraordinary to you.

Let's examine probability as it applies to the sport of kings. Just before the race begins, each horse/jockey combination is assigned a probability of winning. Actually, each pair is assigned odds of winning, but odds may be converted to probabilities and vice versa. If the tote board lists a horse at "4:1", it's a "four to one favorite" meaning the horse has one chance in four of winning the race. Using simple conversion,

$$1 \text{ in } 4 = \frac{1}{4} = .25$$

we can say the horse has a probability of winning or being first of .25, and a probability of not winning—being second, third, ...—of .75. Because winning and not winning represents "all states of nature" in our example, these two probabilities add up to 1.00; in other words, the horse will either come in first or it won't.

Let's create a simple race with four horses,

Horse	Probability of Winning
A	.50
B	.30
C	.15
D	.05
	1.00

noting that the probabilities add up to 1.00. In other words, we're only considering two conditions: each horse either wins or it doesn't. We're not concerned about whether a horse comes in second (places), third (shows) or fourth; winning and not winning are the only states recognized in our race.

When the race is over and the favorite, horse A, wins, we can update our tote board using standard probability theory:

Horse	Probability of Winning
A	1.00
B	0
C	0
D	0

This shows us that, when something occurs—horse A wins—that something's probability becomes absolute or one. The other probabilities that didn't occur—horses B, C, and D didn't win are all given values of zero.

Linear probability theory says that when something probabilistic manifests or becomes real, the probability shifts from a value less than 1.0 to 1.0, and all other probabilities become zero. This is a handy uniplanar convention, but far from what occurs multi-dimensionally.

Probabilities and Multidimensionality

Viewing our horse race multidimensionally, we can say that because there are an infinite number of realities, there are multiple realities in which our race is run. In addition to each horse—A, B, C, D—winning in one reality, there are other realities in which the overall order of horses crossing the finish line varies. For example, there are 6 combinations of the four horses which could all result in a win for A:

ABCD

ABDC

ACBD

ACDB

ADBC

ADCB

However, if we ignore the variations and look only at the winners, multidimensionally *each* horse has a probability of winning equal to one. That is, multidimensionally there's always some reality where each horse wins the race.

This isn't as alien an idea as it might first appear. In many political campaigns, various candidates run to make one sort of a statement or another. They may wish to be the first black, woman, or homosexual to get a majority vote in a particular county in a gubernatorial election, for example. Although they don't win what the majority consider the real race (i.e., they don't become governor of the state), if they win in that one county, that probability may be much more real and meaningful than the one recognized by the general population.

Getting back to our reality in which horse A wins, is it possible to know in advance and from a multidimensional standpoint that A will come first? Obviously those people who bet on A either consciously, subconsciously, or unconsciously know their choice is the proper one. Why doesn't everyone bet on A? For several reasons:

1. If *everyone* bet on A, A's probability of winning would rise to 1.0 and the best a bettor could hope for is to break even—that is, get his or her money back, but no more.

   If their purpose is to make a statement regarding their faith in A's ability to win, it wouldn't make any difference. However, if their purpose is to make money, it wouldn't make sense to bet on a sure thing. These people would either not bet at all or select a prospect who's odds offered greater rewards if the horse wins.

2. Some people distrust their instincts in favor of a more "scientific" approach and therefore bet based on track conditions, weather, the jockey's weight, the horse's past record.

3. The race is a co-created event and therefore some people want to experience losing. They wouldn't bet on A even if A's competition were a blind, three-legged thirty-year-old plug.

In our example, horse A wins because of its own ability *and* because the other horses contributed to its success. Obviously, if there were no other participants, there wouldn't even be a race. Horse A and its rider won because they had the most direct goal; their purpose was the clearest and strongest and they had the least doubts—or possibly none at all—about winning.

What about all those winners who say "I never thought it could happen to me." That may be true, but in a competition where others' feelings of not winning are stronger, such a person could win by probabilistic default. Because nobody believed they could win, the individual having the least feelings about losing became the most probable winner. An intuitive awareness that "someone has to win" is what keeps many bettors betting regardless of the odds. Although the odds for horses, lotteries or blackjack games may vary with the number of participants, for the individual bettor the probability is always 50/50. "I either win or I don't."

Probability in Our Lives

Think of how probability rules our lives. It exists in our basic language with words like "should" and "might". We wonder all the time what will happen to us tomorrow, next week, next year. Is it possible to be more clairvoyant (literally, to see clearly)? Can we increase the likelihood of our chosen probabilities becoming real? Let's look at an interesting phenomenon.

It's fairly well known that as we grow older, certain memories from our younger days become extremely vivid and clear compared to more recent events. Why is this? Let's take the case of an 60-year-old woman who has a heightened awareness of events that occurred when she was 20. Because time in this reality must also be rotational if our rotational theory holds true, our sexagenarian's life cycle can be envisioned as,

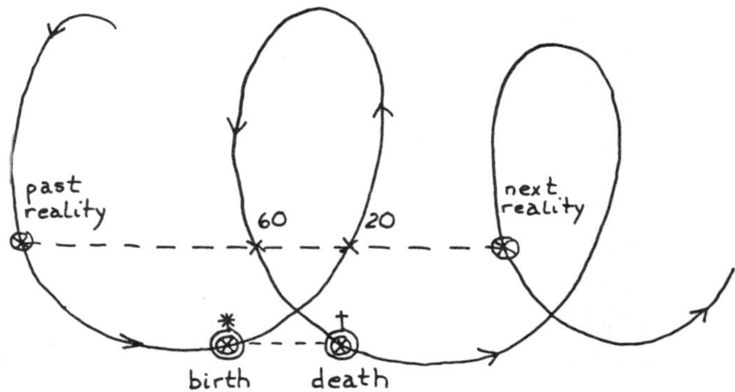

remembering from *Primer* that birth or entrance into this reality and death or leaving it aren't instantaneous events, but rather have a certain "width" to them. We can see in the spiral that the points labeled 20 and 60 are relatively close together. That is, certain times in our life we're closer to other times in our past or future. When our present is more aligned with a point in our probable real past, our memories are more vivid of that past. When we're more aligned with a point in our probable futures, we're more clairvoyant. However, because the twenty-year-old woman seeing herself at sixty has no conscious recognition of whom she's seeing, she's more likely to disregard the awareness of this future probable self. On the other hand, the sixty-year-old is familiar with her twenty-year-old self and immediately recognizes it whereas her alignment with her future self may be incomprehensible and is therefore ignored.

If that's the case, then at the halfway point in life our awareness of the past and our clairvoyance are quite limited resulting in the familiar mid-life crisis. For example, if Joe Brady's life span is eighty years, when he reaches forty he's at the "peak" of his cycle:

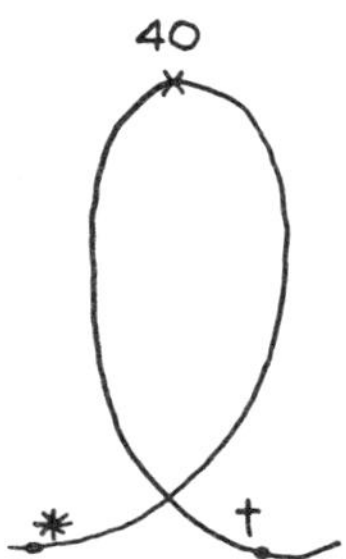

As far as the probabilities he's aligned with in other parts of the cycle, it's essentially more of the same. Unlike our sixty-year-old who's aligned with her twenty-year-old and a self in another reality which can be the source of exciting new thoughts, emotions and experiences, Joe is temporarily trapped within a very narrow past and future. Is it any wonder why at such times, cut off from the stimulation of other selves in other different realities, many sink into depression, search frantically for some meaning in their lives, or strike out on a new course altogether?

Altering Probabilities

The next question is: Can we alter probabilities to suit our needs? The answer is an unqualified, "Yes". We can do this by altering our mental concept of probability. Let's use a common probability, one that nearly all of us have tried at one time or another. "If I were wealthy and healthy, then I'd be happy." In other words, we're saying we believe both health and wealth are necessary for us to be at peace or at one with ourselves and all around us. However, how many of us assign a high probability to being wealthy? Not many. Suppose a friend asks you what your chances of being happy are and you feel you have a one out of ten chance (or 0.1 probability) of having the wealth you desire which is part of your definition of happiness. However, your family is notorious for their excellent health so you assign that prerequisite to your happiness a probability of ten out of ten or 1.00. The equation describing the realization of your goal is

$$(P_w)(P_h) = P_1$$

The probability of being wealthy (P_w) is multiplied by the probability of being healthy (P_h) to produce the real state, P_1. Substituting our probable values

$$(0.1)(1.0) = 0.1$$

we see that by assigning wealth a 0.1 probability we decrease the probability of achieving happiness to a meager one in ten also.

Can we increase P_w to 1.0 and make the probability of being wealthy a sure thing? Yes: We do it in our minds, the *only* place it can be done. It's only in the mind that we define or value our probabilities. For example, suppose you decide you're chances of being wealthy are one in ten. Furthermore, you believe that even if you were wealthy, the probability of your being healthy was only 60% or 0.6, by your *own* definition. What's the probability of you getting what you want—wealth and health?

$$(P_w)(P_h) = P_1$$
$$(0.1)(0.6) = P_1 = 0.06$$

or about one chance in 16! Naturally, if you define the probability of your being healthy or wealthy as zero, the probability of your achieving happiness is also zero.

Because we each create our own realities, the probabilities we assign have a tremendous effect on our behavior. None of us likes to be wrong. Therefore, if we believe "I can never be happy here", the only way we can possibly be happy is if we admit our basic belief about our happiness is wrong. For many it's much easier to maintain a miserable reality than make the necessary belief changes to permit them to be happy.

Are we fooling ourselves by arbitrarily assigning the probability of a sure thing to something we want? Not at all. Because we each create our own reality, we can determine what's important to us and assign it any probability we want. Unfortunately there's a tendency to fall into linear thinking, using past "evidence" to govern our beliefs about future probabilities rather than considering all the probabilities that are available to us right now. For example, if Judy Colson believes "I can't be happy here" because every major appliance and structure has broken down in her home since she bought it six months ago, she immediately assigns her happy state a very low probability. Because she's created a low probability of being happy based on her past experience, each new event is viewed in the same light. The repaired washer could fail again or the new stove break down just like the old one. Instead of seeing the appliances as having their own set of probabilities here and now among which failure or breakdown is one of many courses available, Judy only recognizes *one* negative probability based on her past experience. It's the difference between assigning the washer a probability of failing of 0.001 versus 1.00. By seeing it as it exists in the present, failure is only one of a thousand things that could happen to her washer; in terms of past experiences, it's the *only* thing that can occur. Comparing the two, which orientation is more likely to create a bum washer; which one *needs* a bum washer to maintain the reality in which Judy can't be happy?

If we make our reality the way we do because we want certain experiences, then why not create it the way we want it to be? We do: However, many times we choose not to acknowledge that's what we're doing. "You mean, I *want* to be poor? I *want* to be a fat failure?" Evidently, or you wouldn't see yourself that way. As we said earlier, we often prefer to use the past or observed experiences in the past of others to create excuses why we can't do something in the present or the future. Let's go back to the wealth idea again and assume you want to be wealthy. Looking at your past, you were either wealthy or you weren't. A traditional linear thinker uses either past condition as a way to limit present or future performance.

CASE 1: "I was wealthy once but lost it and would just lose it again."

CASE 2: "I have always been poor and have no hope of ever changing that."

Probability and the Streaming Effect

However, the past doesn't determine the future. This has relevance for us as individuals, for the $_7\alpha$, and for our future human development whether we talk about new energy systems or the enlightenment of the human race. If there are things and events we want, we assign them a probability of one and then all energy, both uniplanar and multidimensional, is brought to bear. And once it starts, it continues. This is the same streaming effect mentioned before and it works with probabilities as well as any other phenomena. Similarly, if we assign something a low probability the chances of it gaining sufficient energy to initiate a streaming effect are equally low.

One of the simplest examples of streaming is a siphon which almost seems to defy gravity the first time we see it operating. Given a tank of liquid we can drain it using only a tube or siphon as long as we fill the tube with liquid first. In the system pictured, liquid enters the tube at level A, reaches a height above the tank at B, and then discharges at C. The critical factors in getting the water to move from one level to another are that the tube must be full and, of course, C must be below A. In other words, to initiate

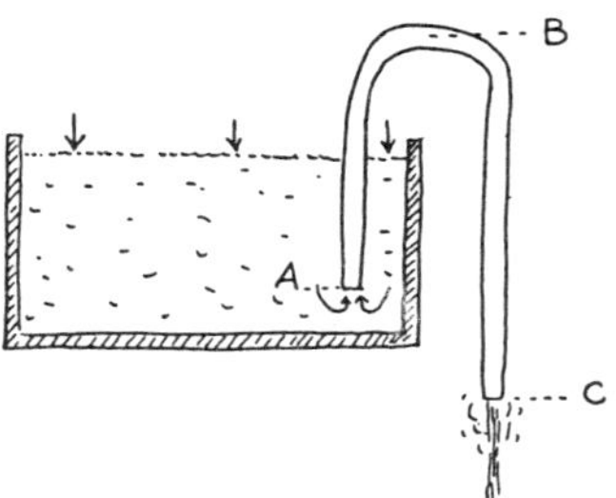

a streaming effect from one level or dimension to another, we must be totally sure what we want right now. If we're not sure, if the tube has pockets of air, the system won't work. As long as the tube is full we can use a very tiny tube to empty a huge vat of water; if the tube isn't full, even the largest one is useless. In such a way, assigning something a probability of 1.0 functions as a filled tube bringing energy from other levels to create the desired reality.

New car owners often become aware of the streaming effect via a completely different mechanism. It's a common experience to buy a new car and then suddenly notice all the cars just like it on the road. They appear almost like magic and we catch ourselves asking, "Where did they all come from? They weren't here yesterday." The explanation?

- We create our own reality.
- We get what we concentrate on.

When one of us was with IBM, a picture of the then-chairman of the board, Thomas Watson, Jr., wearing a unique black and yellow striped tie showed up in a national magazine. The next week it seemed as though every executive in the company was wearing an identical neckpiece. If a popular rock star appears in a certain unique outfit or hair style, it isn't long before thousands of others are copying that style in one way or another.

It's one thing to recognize streaming in objects or events we can perceive, but how does it work in something so unreal as probabilities? Let's return to the races for the answer. Most observers see the finish line linearly:

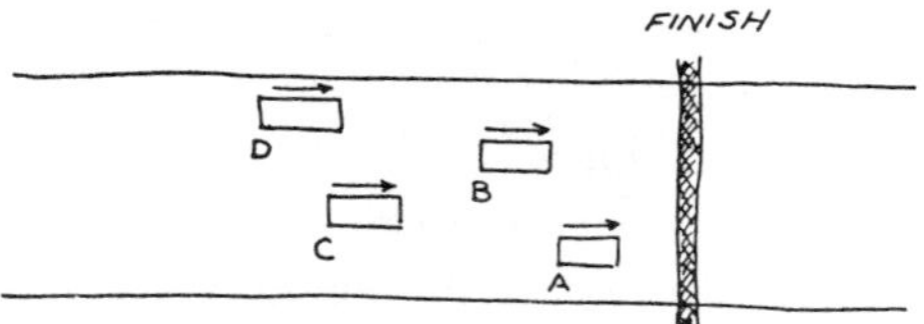

Horse A wins, B comes in second, then C and D, third and fourth respectively. What happens if, instead of linear analysis, we look on the race as occurring outside of time? If time didn't exist, our race would look like this:

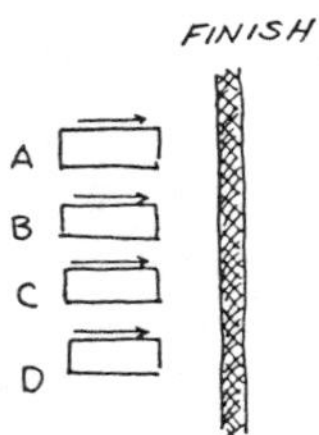

Without time there would be no first, second, third or fourth place horse because those positions are a function of time. If there were no time, there would be no such thing as a faster or slower animal.

Now let's create a finish "line" which is solid except for one opening.

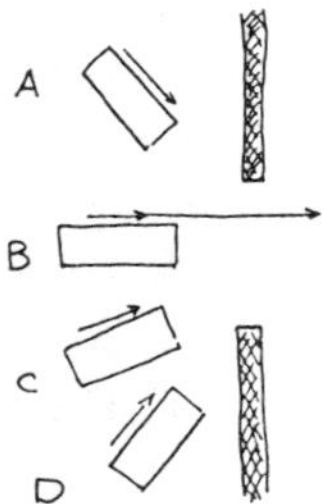

B makes it through the opening and all others (A, C, D) follow. If we define winning as making it through the opening, then the probability of them all winning from a multidimensional standpoint is 1.0. That's why rotationally it doesn't matter whether we win or lose, but how we play the game.

Let's look at another example of the multidimensional probability effect. Instructors in management training often use a game similar to the child's game of ring toss to help trainees assess their willingness to take risks. Each person is given four rings to toss on a peg and may stand as close to or far away from the peg as he or she wishes. The group leader asks each participant to estimate in advance how many ringers they expect to make. The average player stands 7 feet away and is confident of making two ringers.

An interesting phenomenon occurs when a person has only a single ring left and has yet to make a ringer. "Logic" says they should simply throw the last ring away because there's no chance of winning. However, most people will carefully throw that last ring hoping to make a ringer. Even though the real probability of winning the game is already zero, we intuitively view each toss individually and assign it its own probability. Although the whole game may involve the probability of getting two ringers, it's made up of four separate "games" each with their own probability. For each ring tossed, a different set of variables come into play. The person who's made two ringers views the probable results of the third toss differently from the player who has yet to score a ringer. Although from the trainer's linear point of view, there's only one game—the players either get two ringers or they don't—the players are playing that game as well as four separate ones within it.

Is it still hard to think about multiple simultaneous probabilities all equalling 1.0? Remember Agatha Christie's *Murder On the Orient Express* with the redoubtable Belgian detective Hercule Poirot? A particularly nasty man is murdered on the famous train and it's Poirot's job to find the perpetrator. The audience is both entranced and mystified because it seems that each of the 12 suspects all had the necessary combination—means, motive, and opportunity—to do the terrible deed. In the end we discover *all* the suspects stabbed the victim and Poirot is led to the conclusion that if all are guilty, then none are.

Multidimensional Probability and Energy

Hopefully at this point you're not so confused about the concept of multidimensional probability you don't care how it affects our energy principles. It's an extremely important concept and well worth any effort it takes to understand because it shows how what we believe to be all there is may only be a small fraction of all that's available.

Remember our bell-shaped curve in chapter nine:

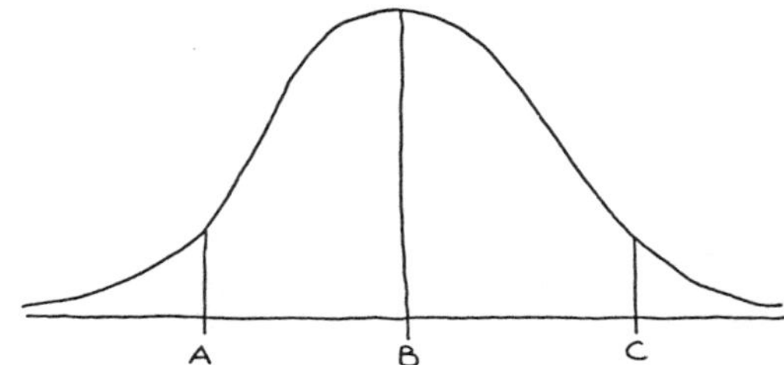

This graph represents all linear probabilities that can occur. Statisticians say the *area* under the curve represents all states of nature and equals 1.0. In other words, if the graph represents the height of all black Canadian females then the average at B might be 5 feet 2 inches. A point at A might represent the number of black female Canadians who are 4 feet 9 inches and point C those who are 5 feet 7 inches. But the total curve represents all Canadian black women and therefore its area must by definition be 1.0.

Now let's move our graph into the multidimensional arena and see what happens.

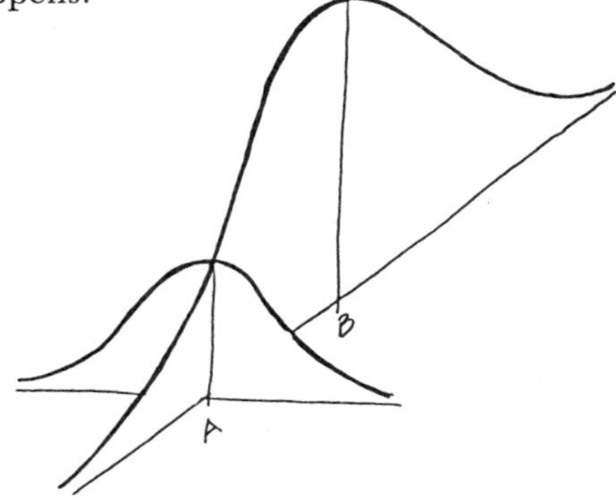

Suddenly our black, female Canadians who are 4 feet 9 have a "world" of their own. If they choose to perceive themselves as "all there is" they may form their own society. To them, anyone who isn't black, female, Canadian or within their height definitions (between 4 feet 7 inches and 4 feet 11 inches, for example) may be viewed as outside their existence or reality.

What do black Canadian women have to do with our energy discussion? Because what holds for any must hold for all, we should be able to apply this concept to a typical energy reaction such as nuclear fission.

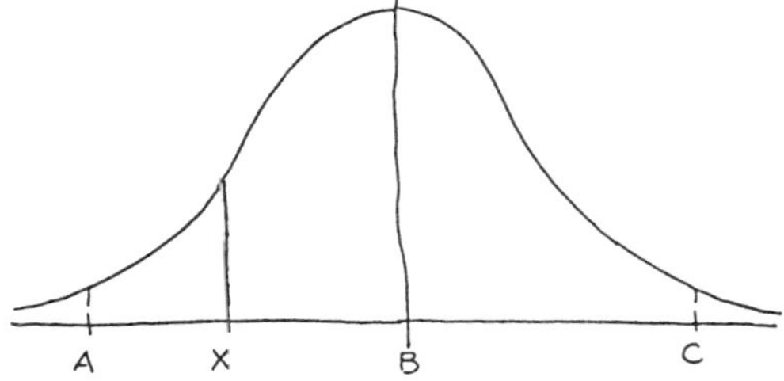

At A, the reaction begins. It increases in intensity through X, reaching a maximum energy "output" at B, and then declines to point C where the reaction is seen to die out. However, because this is a uniplanar probability curve, it doesn't reflect all that's going on at any and all points; it only represents part of the total game that's being played, only one parameter of a much more complex population. For example, if we tap the multidimensional probabilities at point X, we see

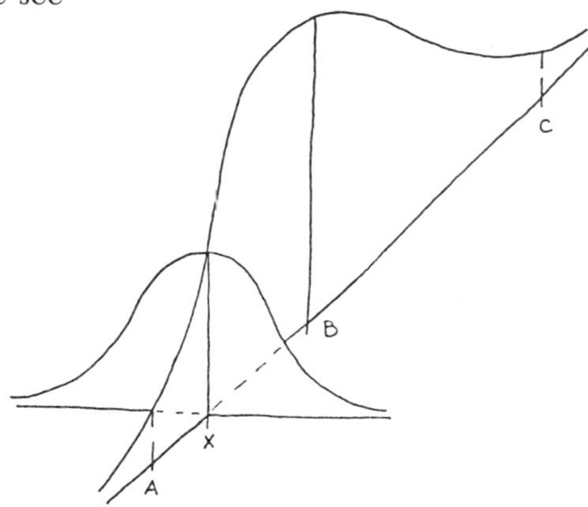

and, if we encourage our fission reaction to produce all that's multidimensionally possible at each point in its linear progression, we get

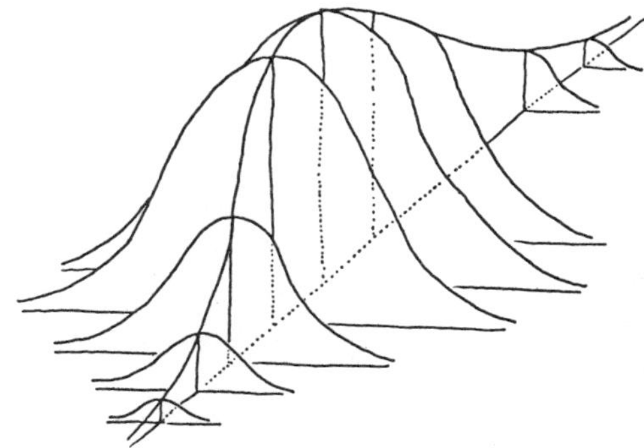

Whereas the uniplanar view only acknowledges the existence of energy production in one form in one plane, the multidimensional approach shows us there are many more forms and planes available. Given such an approach to probabilistic reactions, whether it's burning wood, fissioning a U-235 nucleus, or the fission-fusion reaction of inherent light and water, all of this energy can be collected and utilized.

Any discussion of probabilistic theory would be incomplete if we didn't acknowledge the fact that all such theories must themselves be probable. We offer this as one explanation but not the *only* explanation. As we've seen, there are many explanations of the same thing and this is the way it must be. If there were only *one* explanation, then all must subscribe to it. Without multiple probabilities, we would have no choices and would therefore all be identical. If we all looked, thought and acted the same way, we would have no meaning as individuals. If we all subscribed to exactly the same thing, then all probabilities would cease to exist—and so would we.

Having mastered the complex concept of multidimensional probabilities, let's see whether what we usually think of as energy is really energy—or something else.

15 | The Eleventh Principle

The Eleventh Principle: Any energy collected in a form "above" the level of the $_7\alpha$ isn't energy, it's an energy/mass combination.

If all is composed of $_7\alpha$, how can anything be "above" the level of the $_7\alpha$? Wouldn't any change of state have to affect the $_7\alpha$? Indeed it would and does, but this principle concerns itself with the *origin* of that change rather than its *effect*. Therefore, it's critical we define what we mean by $_7\alpha$-level energy creation or collection because all is $_7\alpha$, and energy as previously defined can neither be created nor destroyed.

The pure energy form of the $_7\alpha$ is the perpendicular pair; therefore all pure energy can be broken down into successively smaller and smaller units comprised of multiples of two. If the mathematics of a reaction, whether the atomic weights or voltage ratios can't ultimately be reduced to 2^x and finally to 2^1 and 2^0 or 1, the sequence is impure: It does not produce pure energy. Obviously we have one sequence ($2^x \ldots 2^2, 2^1, 2^0$) which occurs when we *convert* mass to energy and its reverse when we *collect* energy itself because we begin with a single $_7\alpha$ (2^0), then a pair (2^1), then more pairs (2^x).

When we convert mass to energy properly we essentially align the $_7\alpha$ and collect inherent pairs. Pictorially, the process is from

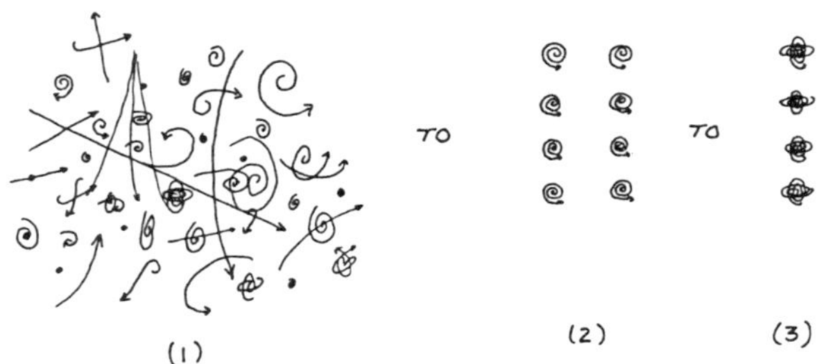

First the incoherent $_7\alpha$ (1) are singularly aligned (2) and then those of equal V_T further aligned as perpendicular pairs (3). Were mass to remain the hodge-podge of $_7\alpha$ shown in (1) above, the prospect of creating inherent pairs from this chaos would be overwhelming. Under such circumstances, we could be tempted to merely blast our way into the mass in hopes of "catching" some purer forms upon impact. However when we understand the most basic rotational physics, we see that creating alignment isn't that difficult.

Creating Alignment

Mass alignment can originate from without or within. This doesn't mean we can force a mass to align; whether alignment occurs or not is a choice of each $_7\alpha$ within the mass. However because all is connected, any change in the environment surrounding the mass creates probabilities which may favor some changes more than others. For example, in a hot room you might choose to remove your jacket. The increased temperature didn't *force* you to disrobe; however it did influence the probability of your doing so.

When we enhance the alignment of a total mass, we create an environment attuned to that mass. Obviously there's already some "attunement" present or the mass would be unstable. However, what we're referring to is the difference between the attunement of an orchestra and a single instrument versus that between a tuning fork and one note from one instrument. In other words, rotational theory wants to fine tune and synchronize *both* the immediate environment and the mass.

One way to accomplish this is to match V_T or more simply, V_S or V_L. In essence this is what cyclotrons and other accelerators do. However, because the velocities being matched are based on the calculated values of sub-mass "particles", they exceed natural environment-mass interactions and therefore produce abnormal and unnatural intermediate states. Consequently as a source of either natural alignment or energy collection any form of acceleration, including our previously-discussed chain reaction, is nonproductive.

If we want to create alignment from the outside in, we must

begin from the outside. If we begin with something that's unstable, then what we produce will also be unstable. Driving a defective vehicle at high speeds doesn't decrease its defectiveness any more than bombarding an unsafe radioactive mass can ever produce safe energy. Like begets like; we get out what we put in.

To align something from the outside in, we must first match the environment and the mass. One simple way to do this is to encase the mass with an envelope of material having a compatible but lesser atomic weight. In such a way the system is stable but the higher energy state of the envelope aligns the greater mass state simply by diffusion:

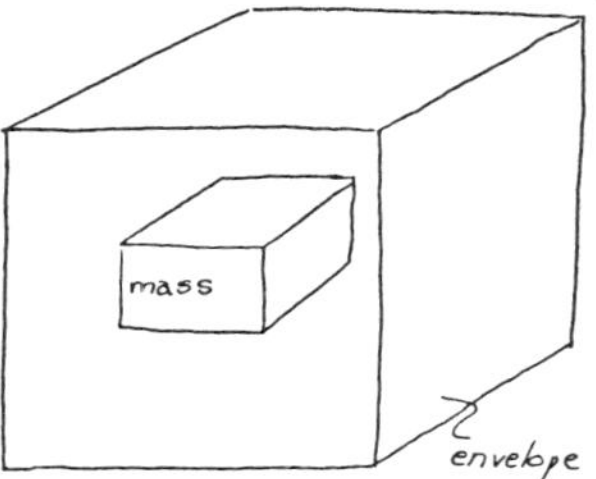

Looking at a side view, microscopically we see

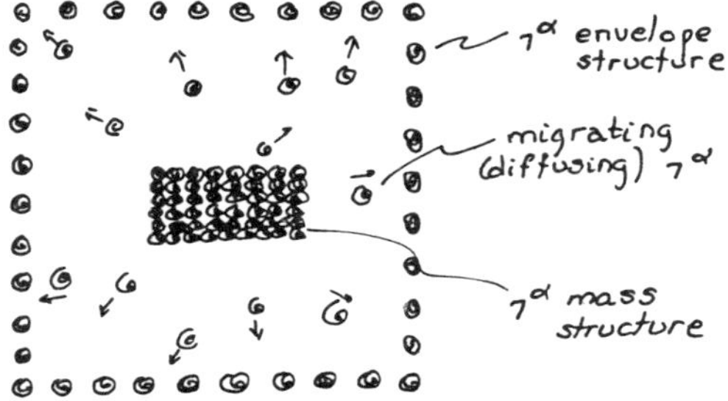

As we can see in the microscopic view, the $_7\alpha$ in the interphase between the two layers have much more room. What happens to $_7\alpha$ that have more room? They stretch out and become more energy-like. Meanwhile as diffusion occurs, our mass experiences internal changes in response to the migration and changes occurring in its surface $_7\alpha$. If the mass is totally suspended within the envelope, there's uniform migration of $_7\alpha$ from a central point to the surface of the mass. The smaller, denser higher spin $_7\alpha$ are the last to migrate because the $_7\alpha$ now between our two original forms tend to attract those most like themselves.

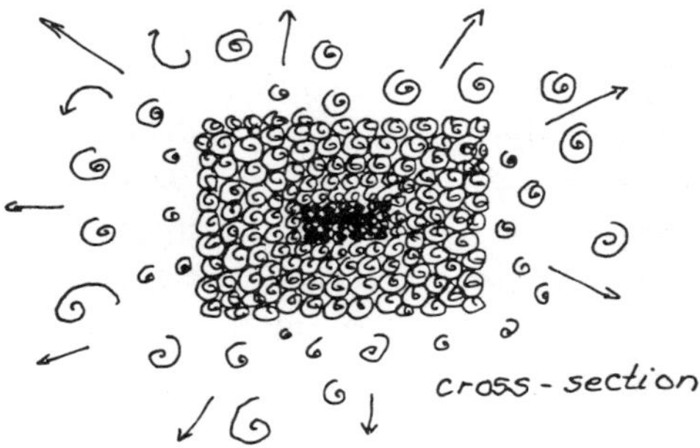

Because the first to migrate had the most space, they were the most linear and they then attracted those that were most like themselves.

Remember in school when the phys. ed. teacher had the class form teams for some competitive game? If the game involved a lot of running or energy, the captains always chose the fastest individuals first. We can even say that the presence of the high energy teacher served as the envelope; because of his or her preferences, the game selected was a high energy one. Chances are the instructor even picked the two fastest runners in the class to be team captains, and they in turn first chose those most like themselves. Surely we all remember our discomfort at either observing or being the last to be chosen—that person so different, so lackadaisical and un-coordinated compared to the teacher's or captain's great ability, it was almost painful to bear.

If our system has no way to rid itself of these migrating $_7\alpha$, eventually equilibrium is established between envelope and mass and no net flow occurs. However, if we apply a relative vacuum between the two layers once the flow is established, we can easily siphon off these nicely separated, aligned forms.

The problem with either collecting or converting energy at the smallest, paired $_7\alpha$-level is: How do we recognize that level? In terms of present systems and methods, the most obvious characteristic of $_7\alpha$-level conversion is the lack of waste. If we place a highly dense material such as lead or even natural non-radioactive uranium within a vanadium or titanium shell, all we get off is pure energy and all that remains is *unchanged* mass. This occurs because we're using a *surface* reaction, similar to what occurs when water evaporates or converts to the more energy-like vapor state without altering or harming the configuration and properties of that remaining. It should be obvious from our awareness of the natural processes of evaporation, erosion, and just aging that this sort of conversion is relatively slow—too slow, in fact, to fill the vast energy needs of current technological societies. We can resolve this dilemma in one of two ways:

- Change the way we live
- Speed up the natural conversion reaction.

Passive Mass Conversion as an Energy Source

Although our ultimate goal is to make the necessary individual philosophical changes to change an entire species because technological changes at most affect only a few, our immediate energy problems must also be resolved. Unfortunately most of us consider philosophical changes leisure-time activities because they tend to be more complex and multidimensional; rather than just considering the ramifications of each singular point in a linear reaction, philosophy delves into the probabilities of all points. While the result is a much more accurate picture of the event, it also takes much more time. What the philosopher does now may have little or no meaning for the present; its value may lie in the changes it proposes for future generations. Or so we believe.

Consequently, when we feel deprived now, it's difficult to consider changes for the betterment of those in the future, even though that may mean a better world for our children. Any real improvement must occur *now*, not in some distant probabilistic future for it to make a difference. We know each one of us can and must take responsibility for changing our energy philosophies but while we're doing that, what can we do to solve the energy problem now?

The best way to solve the energy problem is to accelerate the process of passive conversion of mass to energy in such a way as to be noninvasive or detrimental to the process itself. So often when engineers think of increasing the rate or speed of a system, they think in terms of applying more force to it. This is accomplished by

- Supplying more fuel to an internal combustion engine.
- Increasing the pressure in a water system.
- Turning a turbine faster.
- Adding heat to a chemical reaction.

In such ways it's believed output or production is increased; however, such action invariably creates wear and tear, friction, fatigue and tension as well as waste.

A better way to increase output is to increase the rate of product removal. This initiates the streaming effect which is then passively increased or maintained by the system itself. Although this is relatively easy to envision in terms of our enveloped mass where the application of a simple vacuum between the layers speeds the conversion, how do we remove the "product" of motion? If a system such as a car is designed to cover distance, how do we remove the distance-product to keep the system going? Obviously any time a moving system moves, distance is (re)moved. So like any mass-energy conversion, it has a product, only in this case it's harder to perceive. If we have mass sitting in a chamber and it appears to get smaller, we can easily recognize a conversion of one form to another, even though we may not understand the specific process. On the other hand, when our mass goes from point A to point B and the change is the distance AB—and the mass appears to be exactly the same at points A and B—it's much harder to think in terms of any collectible energy form being created. Without the presence of a recognizable form, we can't remove it and apply streaming mechanics to the system.

Mass-Energy Conversion and Diffusion

Are we doomed to increase motion by applying more force *to* the system rather than passively removing motion *from* it? Isn't there some way we can passively increase the speed of a car, for example? We already know simple earth-mass diffusion or gravity "pulls" a car from the top of a hill to the bottom. In terms of energy diffusion, we recognize that if the car is rarefied mass relative to the top of the hill, then its energy-ness must correspondingly be greatest at that point:

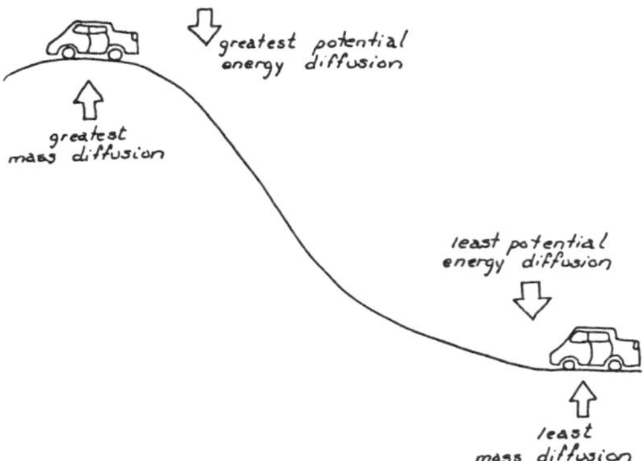

When the car is at rest we have two simultaneous systems at work. First there is more mass beneath the car than in the car itself; the mass of the earth is obviously greater than that of the single vehicle. Therefore, because we know passive diffusion goes from most to least, the higher spin more mass-like $_7\alpha$ migrate from the earth to the resting mass forming a "bond". Secondly, if the car has less mass, it must have more energy when at rest than the earth beneath it. Therefore, in addition to the more mass-like $_7\alpha$ migrating or diffusing towards the car, there are some more energy-like ones migrating away. However because the air around the car contains a greater number of the more energy-like $_7\alpha$, the air tends to replenish those lost from the car to the earth—leaving the net energy content relatively unchanged.

Furthermore, because rotationally energy *is* motion, we see no greatest energy state as long as the car is at rest. Potential energy has no meaning because there is no motion. Although physicists use the term potential energy to describe that present within an unmoving object by virtue of its position, the fact remains that potential energy is like a promise: We have no real proof any energy exists until the system starts moving.

At the top and bottom of our incline we have the least and most mass states *relative* to the earth, so somewhere between we must have the most and least true or kinetic energy states. Bear in mind we're speaking in terms of the car's relationship to its environment, not the mass-to-energy ratio within the car itself. We're looking at motion as the result of the interaction between the earth-car system rather than a singular change affecting only the car.

The instant the car moves, our diffusion gradients change completely. The mass diffusion bond holding the car to the earth is broken and the vehicle now exists in a state of greater energy-ness. Now the greatest diffusion occurs between the $_7\alpha$ of the car and the atmosphere. Because we know the latter has a greater concentration of more energy-like $_7\alpha$, these diffuse into the vehicle. Simultaneously the greater mass-like $_7\alpha$ of the car diffuse into the atmosphere rendering it even more energy-like.

Determining Maximum Mass and Energy States

Returning to our hill, the car achieves its most energy-like state at a point 4/5 of the distance from its highest *resting* point. This "4/5 point" can also be found using standard Newtonian physics. Consider an object about to fall from a height, h:

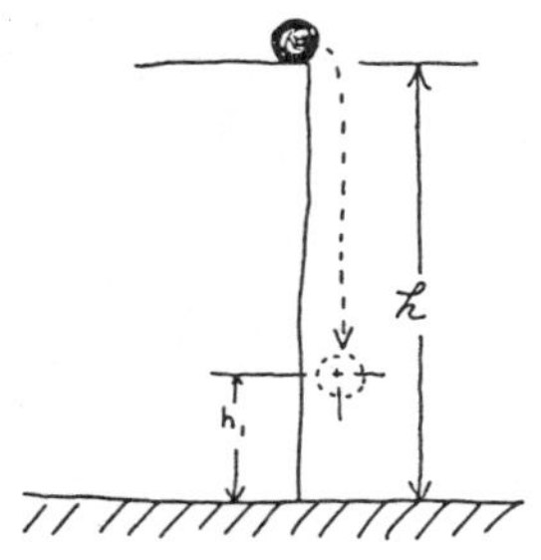

If we know its potential energy, mgh_1, where m is the object's mass, h_1 its height above the ground, and g, the gravitational constant of 9.8 m/sec^2, and its kinetic energy of 1/2 mv^2 where v is the object's average velocity through the distance $(h - h_1)$, we can equate potential and kinetic energies,

$$mgh_1 = \frac{1}{2} mv^2$$

and solve for that height where the two are equal. We find that point is 4/5 "down" from the total height, h. So we can locate the point of our car's greatest energy state:

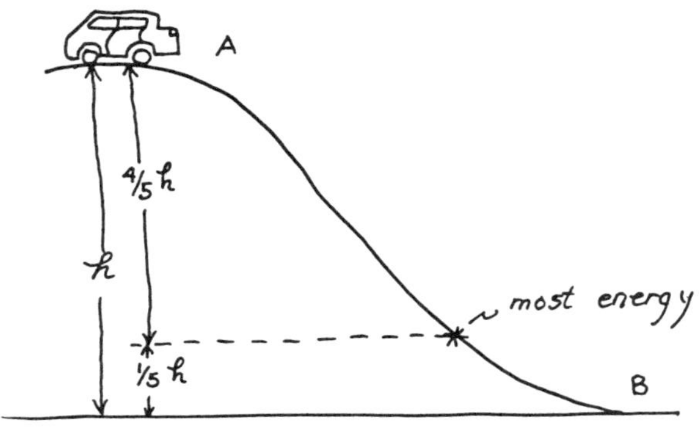

Because the two resting points are defined as (relative) car mass, we also know they must be states of (relative) low or no energy—again ignoring potential energy since it's not "real" energy. Thus when we graph our car in terms of mass and energy relative to distance, we see

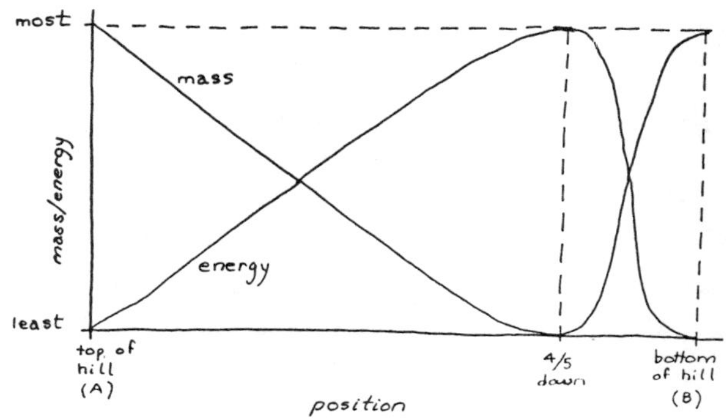

This graph can teach us a great deal about mass-energy conversion. First of all, like all things in rotational theory it isn't limited in its use to cars careening down hills or apples falling out of trees, striking certain physicists contemplating the nature of gravity. All we need to know is the direction of either mass or energy *change* and we can complete the graph. If we want to know what happens when an object moves *up* a hill, we merely reverse the graph. Obviously if we put x units of mass into a system, we can get x units out—not at the *end*, but rather at a point 1/5 from the end or 4/5 from the beginning.

Let's briefly consider other probable meanings of the graph. Think of all the references to things getting worse before they get better: the ever-present ultimate darkness before dawn. Time and time again mankind analogically refers to this point of maximum change and energy-ness that is absolutely critical if any distance is to be covered, any lasting change to be made. Yet time and time again, we ignore the obvious and even denigrate this point of maximum energy-ness equating it to our darkest hour. As even Newtonian physics tells us, at the point of greatest energy-ness our awareness of our mass-ness must be least; so obviously when we experience periods of great change, we're bound to be disoriented at first. When we're moving we look and feel different from when we're standing still; similarly all else looks different to us. Because we expect everything else to remain the same, we often become frightened by the idea of change and choose to remain stationary. Once we know this is a normal part of any change we can accept it, and it becomes a different rather than negative experience. Furthermore, recognizing that period of greatest energy, change and disorientation is well over half way (4/5 of the way, in fact) towards its mass manifestation, we can bear any discomfort more easily.

Now let's convert our car to a bouncing ball and expand the concept:

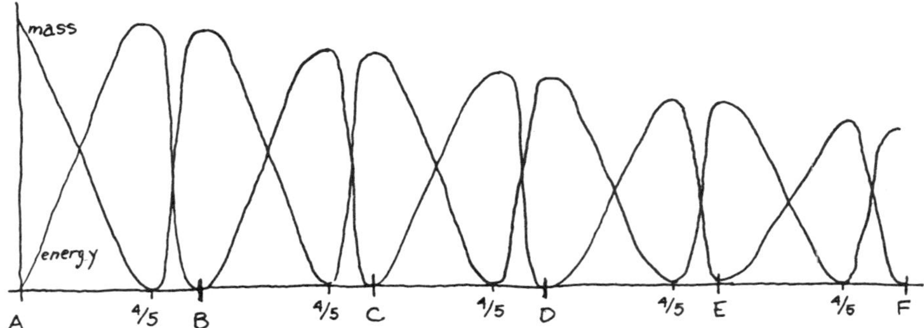

Here, the "4/5" notation refers to points where maximum energy-ness and minimum mass-ness occur. As we can see, the curves dissipate at a predictable and calculable rate—much like that of a damped sine wave. If we "pull" the curve slightly, the pattern forms the familiar double helix:

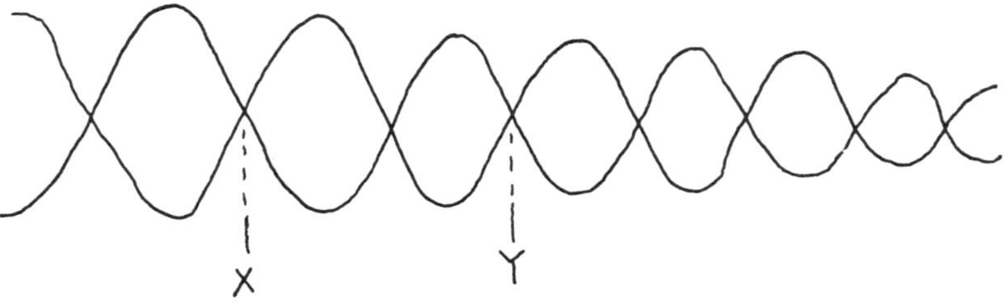

If we examine one set of "loops"—the area between X and Y, for example—what do we learn about the process of energy-mass conversion?

- For each unit conversion of mass to energy, there's a maximum and minimum value achieved for each.
- Between each maximum and minimum state, there's a point where mass and energy are equal.

Therefore we can assign values to our graph in the following manner: Because each unit conversion per unit time represents the maximum of one component or *all* of it that is available versus the *least* of the other, we may say that one's zero is the other's *infinite:*

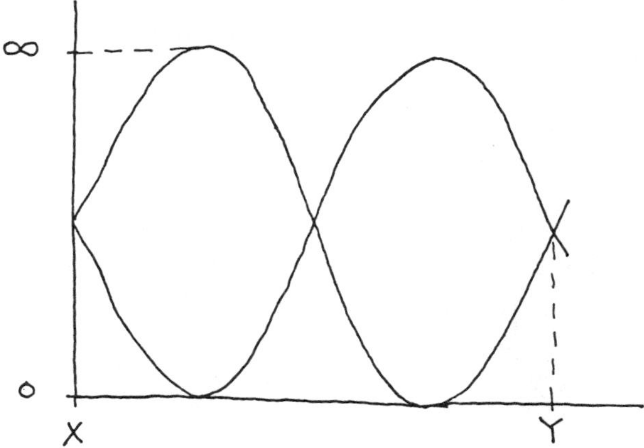

In essence what we're doing is converting specific uniplanar values to rotational ones that enable us to relate all kinds of different reactions. It's not important that the maximum energy state produces x BTUs of heat or y volts of electricity; nor do we care that the maximum mass state is 2 grams. All we need to know is that when the mass is least (zero), energy is maximum (infinite) in that system: Thus each system has its own zero and its own infinity.

Flow, the streaming effect, thermodynamic transfer and the

motion of falling bodies are all forms of diffusion. In all cases there's a flow of mass and a corresponding flow of energy. The reason both water and electrical current follow the path of least resistance is because the more space, the more the $_7\alpha$ can stretch out. The more it stretches out, the more energy-like it becomes; the more energy, the more motion, the more motion, the faster the flow.

Balance and Conversion

From this we can see how a conversion reaction is really composed of *two* reactions, both adhering to the same principles. First we have the reaction between the mass and energy components of the system and secondly that between the system and any mass and/or energy in the environment. To determine whether an object first begins moving in response to its inner balance or the environment launches us into the familiar chicken-egg dilemma. However, we do know from Newtonian mechanics that an object placed in a vacuum* or outer space tends to remain stationary unless acted on or, more correctly, unless acting *in response to* another mass or force. However, we also know that any object constantly exchanges $_7\alpha$ with its environment. Because there's so little mass in a vacuum, the mass-like $_7\alpha$ will tend to migrate away from the object and the high energy imperceptible forms present in the vacuum will migrate towards it. Depending on the degree of vacuum and the amount of environmental energy, the mass of the object (actually its density) may be greatly reduced under such conditions although it may maintain its *perceptual* integrity.

The mere fact that a small force can impart infinite motion to a fixed object in space or a vacuum is a testimony to the more energized form in which that resting state exists. Think of sneaking up behind a highly agitated spouse and gently blowing in her ear or lightly stroking his neck. There are some folks who will leap away with such force it's easy to believe they could run forever. The same occurs in outer space; because the internal mass-energy

---

*A vacuum is traditionally defined as a space devoid of matter; that is, one void of *perceptable* mass or energy.

balance has shifted toward energy-ness, it's much more responsive to outside forces. Unfortunately, in both physics and psychology this is seen as avoidance behavior rather than a natural response.

If a force, A, is applied to a fixed mass, B,

and B then moves in a direction A′,

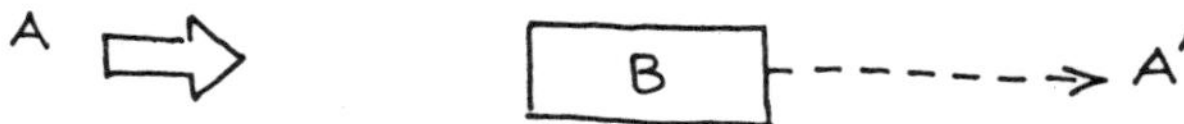

the energy state of B has joined that of A. If A is a strong energy force with little mass component—such as a high wind, for example—B moves in the same direction as A in response to energy diffusion. If force A has a mass as well as an energy component (Clint Eastwood throwing a punch at a villain), B responds to the greater component first. A high-energy fast punch causes the other to reel in the same direction as our hero's fist; however, once the motion dissipates or equalizes, the interaction shifts to the mass arena—the dynamics between mass A (Clint's fist) and mass B (the villain's jaw). If the mass of A exceeds that of B, B is drawn toward A, which then stops B's relative motion away from A.

We can also look at heat transfer in terms of diffusion. A hot object expands because it's more energy-like and therefore has less mass than a cooler one. If it's sitting in a high mass, low energy environment (a hot block of metal on a cool bench), it wants to diffuse in the direction of the greater mass. It doesn't necessarily give up heat energy so much as take back mass. Simultaneously the heat- or energy-poor environment of the bench is quite mass-concentrated compared to the block so it tends to pick up energy and dissipate mass.

Combined Forms and Nuclear Reactions

As we bring this chapter to a close, let's look again at our nuclear reaction. The internal reaction wants to be one with itself. However by forcing it via the application of greater external mass and energy to respond to unnatural gradients, the reaction is always working against itself and therefore highly unstable. Our ball in outer space is constantly giving up $_7\alpha$ in the form of mass and accepting those in the form of energy. Eventually it reaches a point where it's composed of a stable, but ever-changing population of $_7\alpha$ reflecting the equilibrium of the ball with its environment. Because visual perception is a function of energy not mass, we could see the ball at this point but it wouldn't have the "solid" look of mass. Because the system is stable in a state of energy-ness sufficient for perception to occur, that perception will last forever providing this equilibrium isn't destroyed. Furthermore, because we can see the ball we know it's an infinite source of energy. How do we know this? Because the retinas of our eyes function essentially as little energy collectors, with the nerves and brain serving to transport and convert that energy into e-m (thought) or some mass change. In order for us to perceive the ball, it must be emitting energy constantly to stimulate the receptors necessary for perception to occur.

However if someone comes along and says, "I know that ball is emitting pure energy and I want it all now," and subjects the ball to incoherent high heat, pressure, or some other relative energy/mass combination, the ball is destroyed. Rather than creating more energy, the person destroys the natural infinite flow that existed. Because the form creating the ball's mass is so much more energy-like, its $_7\alpha$ so much more open and linear, energy forms such as incoherent heat and pressure function as *mass* to it.

When our low mass, high energy ball is subjected to the relatively greater mass of incoherent heat or pressure, it's literally pulled apart via diffusion mechanics. Doesn't that free the ball's energy to establish flow with that of the heat and pressure? It would, except that the heat and pressure are so incoherent they function as mass and more mass rather than energy and mass. Because the energy form of the heat and pressure is already tied up in the reaction with the ball's mass, there's nothing the pure linear energy recognizes as "like me" to establish flow. What we see as energy from such a reaction is actually a burst of incoherent mass-like forms composed of nonlinear $_7\alpha$ from ball, heat and pressure.

Remember: The streaming effect depends on like attracting like, getting out what we put in. Whenever two systems come together, the likes seek each other out; if there are no likes, that entity becomes waste—regardless of its form. Thus our intermediate $_7\alpha$ forms in the heat, pressure and ball find each other and form an intermediate state we erroneously label energy. All $_7\alpha$ which don't exhibit that form become waste—whether they're the pure linear forms that slip unnoticed into space or the more mass-like ones we recognize as ash or debris.

What if we replace our ball with an unstable mass like U-235? Because now our mass and our mass-like heat and pressure are all unnatural forms, the products and waste of the reaction must be equally unnatural and therefore unstable; like begets like, we get out what we put in. The "energy", like that created in our ball system, is merely a less mass state. The leftover "mass" is a bastard form consisting of trapped pockets of energy within an unstable matrix: radioactive waste. What makes radioactive waste radioactive is the entrapment of unnatural energy in an unnatural mass, energy that wants to get out of there. Unlike such occurrences in nature where such pockets of energy $_7\alpha$ trapped within dense mass can respond to natural energy draws such as cosmic radiation and sun spots and diffuse gradually and harmlessly, the man-made radio-active waste is too unnatural to respond to any natural gradients. Like begets and wants to be with like. Remember what happened to Frankenstein's monster when he discovered he was the only one, the freak, the unloved? When there is no other one "like me" to balance the system, to co-create an at-one state, the system's extremely unstable. Thus our radioactive waste sits and sits, finding so few likes, so few other monsters to form stable pairs with per unit time, the process may take eons.

Wouldn't the process stabilize faster if we didn't shield the radioactive waste? It certainly would. However our problem is the same as Dr. Frankenstein's: When we create unnatural forms, how can we keep them from interacting with and altering, "almost" or "kind of like me" forms as they seek to establish an at-one state? How can we keep the monster from ravaging a fair maiden or strangling a meter reader in the name of love or companionship? How can we keep the aberrant forms from attempting a *pas de deux* with our children's, our dog's, our maple tree's genetic ($_7\alpha$) code? The answer is: We can't. Unless a critical distance or shield between the two is maintained, such a probability always exists. The creation of energy at any but the $_7\alpha$ level not only leaves us with questionable energy, it quite literally produces waste which traps us between a rock and a hard place. If we shield the waste, we protect ourselves but greatly prolong the time necessary for these forms to stabilize. If we don't shield it, we increase the probable number of equally aberrant forms coming in to stabilize those trying to get out, thereby decreasing the time necessary to achieve the at-one state; however, we also increase the probability those aberrant forms may combine with something else. To create or collect any impure energy/mass combination with its minimal usable product and large amounts of unusable and/or highly unstable waste makes this form a monster hardly worth the effort.

In our next chapter we'll use our final principle to show how a lot of little Davids have a lot more to offer than a few big Goliaths.

16 | The Twelfth Principle

The Twelfth Principle: The total energy of any state is the sum of the energy of all its components at any given instant.

To appreciate the full meaning of our final principle, we should both capitalize and underline the word "all" to remind us that energy, like all things, is multidimensional. That being the case, we can't use a system based on uniplanar perception to measure total energy because the total includes the perceived *and* unperceived, the real and unreal. For example, suppose you wanted to know the total number of people living in your neighborhood. Although a count of the houses would give you a reasonably accurate idea of the minimum number of occupants, it wouldn't necessarily be indicative of the total. One large colonial house may be the abode of a singular eccentric while the identical structure across the street could house thirty boisterous engineering students. We can compare our count of the houses to the amount of perceived uniplanar energy and that of the actual residents to the total energy output.

At this point a logical question comes to mind: What difference does it make if the energy isn't real, if it can't be perceived? Obviously it makes no difference at all if real is *absolute*. But it's not; of all things, what's real is the most arbitrary and capricious of all. Because of this it's much easier to describe the sum of all of anything or anyone in terms of simultaneous *and* sequential probabilities. It's the simultaneous manifestation of all probabilities in a single instant which produces REALITY NOW; it's a *sequence* of such points which produce reality in the past, present, and future.

Perception and Probability

Let's look at a simple curve:

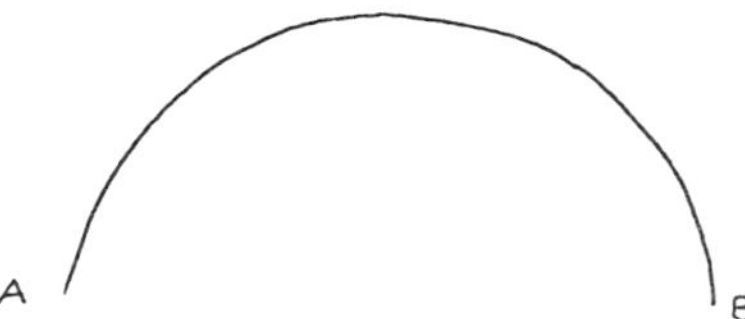

We know this "solid" line is actually made up of dots, much like those comprising a TV picture or newpaper photo.

The solid line represents our working or real view and the dotted one our expanded view. There are times when each view better serves our purposes. If we're interested in the overall shape of the curve, the solid line is preferable; if we're interested in locating a specific point on the curve, the dotted representation is more helpful. Compare this to going on a vacation where your goal is the Florida Everglades versus a sequence of stops culminating in a visit to that area. In the first instance the trip is viewed as a continuous process necessary to reach the goal; in the latter the linear process is perceived as a series of processes each with its own goal.

Now let's shift our reference plane so we can see "beyond" our curved line and its points. From this vantage we discover each point is also composed of an infinite number of points,

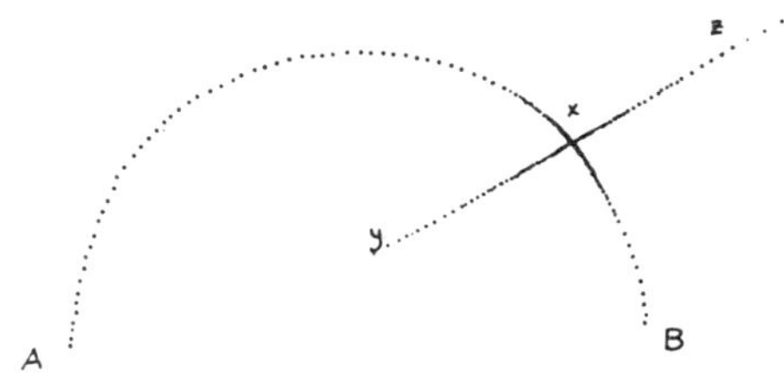

and if we look closely we see there are more points clustered on either "side" of what we perceived as the singular point in the line. Although the line of points stretches infinitely in both directions, our ability to perceive them creates a bell-shaped (Gaussian) distribution

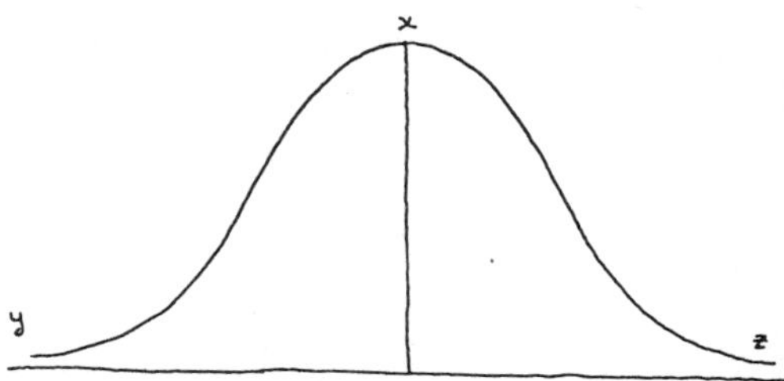

with its peak, x, coinciding with the visible point in the visible line. In other words we now can acknowledge that what we originally recognized as a single point, x, in curve AB has its own unique probability curve, yz. For example, if the Washington Monument were one of the points of interest on our trip to the everglades, it's our main focus in the Washington, D.C. area. Therefore, most of our ideas about Washington center around the monument even though we know the city has other infinite experiences to offer.

Now our awareness has stretched from a single line, to a line of points, to a line of points each composed of an infinite array of points assuming a bell-shaped probability curve. Suppose we want to know how many points are inherent in our original line AB, or how much distance is covered or motion produced when we traverse that line. In our initial view, the total distance is equivalent to the length of the line—the distance between your home and the everglades, for example. In the expanded view we run into difficulty calculating the total distance because we've defined the line as a progression of dots or stops in our trip. To say the distance covered within each stop adds nothing to our calculation of that between our house and the everglades is ludicrous.

Therefore we can see how our definition of the total distance as that between our starting and ending points loses meaning as soon as we make any stops. The more places we pause in our journey to walk around or visit a historical sight or two, the more distance we add to our total. Because we know any line is composed of an infinite number of points, we can say that to experience them all we would have to cover an infinite amount of distance. In other words, if we stopped at every single point between our home in Boston and the everglades, we would cover an infinite amount of distance compared to the 1500 miles covered if we made no stops at all.

To further complicate our problem, the multidimensional perspective expands the distance of each point infinitely in both directions along a line perpendicular to the original or real line. Suddenly, describing the total distance inherent in our simple, real line becomes quite complex! A quick survey of this mess leads us to intuitively recognize that the actual distance covered in any line *must* be infinite. However, because there are measuring devices and computers designed to "prove" our line is only x cm long, we must find some concrete way to prove our intuitions regarding our line's infinite length.

Proving Infinite States Using Time

We already know the concept of infinity is perceptually cumbersome because of our finite orientation. Therefore if we could find some way to reduce our infinite line to smaller, more manageable increments, it would be easier to relate the line to something we know. The most obvious way would be to isolate a single point in the real line and then study it in detail. But how can we isolate a single point of an infinite progression? By introducing time: Time makes the line or any point on it real.

For example, imagine yourself drawing our first simple curve. At any instant during the process there's a part of the curve which is real (drawn) and part which is unreal (not drawn)—places you've already been, those you have yet to see. If we measure the time it takes to make the real line, we know that time, t, divided by infinity (t/∞) equals the time it takes to draw a single point. However because we realize infinity, like energy, is beyond our perception, this doesn't help us much. On the other hand, we can create a smallest-point by dividing the total length or distance by the amount of time it takes to draw it. Even though we recognize your drawing rate may change during the procedure, this value gives us an average-size point. Once we have that real point, we can then begin studying its unreal or multidimensional qualities. For example, if it takes you five hours to drive two hundred miles and you make five stops in that time, by dividing that distance by the time

$$200/5 = 40$$

you can break your linear travel into four packets or "points" of forty miles each. Those of us who keep track of our gas mileage use a variation of this technique. Although we realize we get fewer miles to the gallon in start/stop city traffic versus driving on the open highway, we determine our average "point" by dividing the total number of gallons into the total number of miles covered.

In a similar fashion, we can break down any energy-producing reaction. First, we recognize what we start with, what we end with, and what we see in the middle; then we express these changes as a function of real time. Next we take the "total" real energy and the "total" real time and determine the rate of energy production per unit time. This gives us a workable smallest unit from which we can make our multidimensional calculations.

For example, if the production schedule of a ballet company binds a ballerina to one performance per week and any related rehearsals and post-performance activities, she's free to do as she chooses with the rest of her time. In this case let's say she accomplishes her job in about one day per week, total time. However, she doesn't sit around waiting for the performance and related events the other six days. She involves herself in all kinds of other different, and often equally creative, endeavors.

Now let's suppose she's paid $1,000 a week to perform and, according to her contract, that's all she can do. She must do nothing but that performance and its directly related activities, which we know only take up one day of her time. Her director feels she's worth it, because the performance is a hit. It's more important to him to pay her not to work six days and be sure of her performance on the seventh, than pay her less for one day's work and have her seek other employment the rest of the time.

It's quite conceivable our dancer simply can't be idle for six days. Whether she gets paid for it or not, she'll make use of her creative talents during her free time. When we analyze this system strictly from a business point of view we see

- The employer is paying more than is necessary for what he's getting because he's paying her not to work those other six days in addition to working one day for him.
- Because he sees only one goal, the one performance, he can't see the dancer's other creative talents as an equally marketable and profitable commodity; he sees them only as *threats* to his production.

Consequently he forbids the dancer to do anything but that which he requires—which utilizes only a portion of her potential. She eventually becomes frustrated with this situation—"Any first year ballet student can do what I do"—and quits the company.

In our rather elaborate analogy we see how the desired product determines the director's idea of the time necessary to achieve it. However it doesn't take our skilled dancer the whole week at all, but only one-seventh of it—one day's total time. During the other six days she's capable of manifesting many different skills in many different ways. But because her employer has defined the goal as requiring 7 days of *linear process* to achieve, he can't recognize these events. If he does see them, he sees them in terms of being detrimental to the dancer's ability to perform certain regimented activities only on a certain day.

Time and Nuclear Reactions

Now let's move this into the realm of nuclear activities. The "performers" in a nuclear reaction aren't as lucky as our dancer. They're not allowed to leave; nor are their other, non-linear, activities acknowledged. The nuclear engineer, like our director, sees the linear nuclear reaction as all there is. He or she assumes that whatever changes occur require the entire time interval between when the reaction starts and when it ends. However, we know that interval is composed of an infinite number of points like our line. In order to locate a single point in the reaction, we can use the same process of relating known events to time. Let's define the interval between the first impact of a neutron with a nucleus and the subsequent collision of the resulting pair of nuclei with the next pair of nuclei as a *neutron-second* (n–s). We can measure this same interval in real time and discover our neutron-second is equivalent to 1/10,000 (10^{-4}) second.

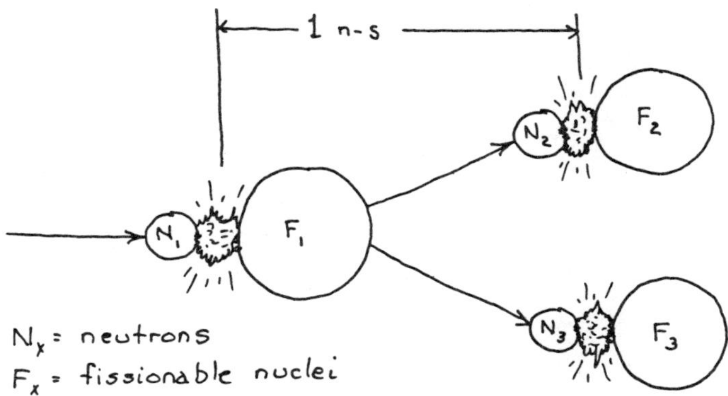

However, because the neutron-second describes the span between neutron "bursts" or impacts regardless where they occur in the linear reaction, it *can't* have an absolute value. Due to the aberrant nature of chain reactions, we know these intervals become much shorter as the reaction proceeds. Therefore, although our first neutron-second is equivalent to 10^{-4} sec. linear time, the second may be decreased in linear time equivalency by a factor of 1:10, the third decreased by 1:100, and so on.

To the observer, then, it appears to take 10^{-4} sec. for one neutron to strike the target and produce more neutrons which then strike mass again. We automatically assume everything occurring in that interval is associated with the process; once the first neutron strikes the target, all is geared toward the release of more neutrons to strike more mass and so on. The view is primarily one of a billiard ball breaking up a mass of balls and then other balls going on to break up other masses. Because nuclear fission is considered primarily a mass-to-energy conversion, the analogy is valid.

However, let's look at this rotationally and multidimensionally. Think of throwing pebbles in a pond; in this case our "pebbles" are tiny, rapidly-spinning whirlpools of water. As soon as they enter the pond, both they and the pond begin to change. It's unlikely our water-pebble will translate or move linearly much, if at all, once it comes in contract with the water. However it will radiate its spin into its immediate surroundings:

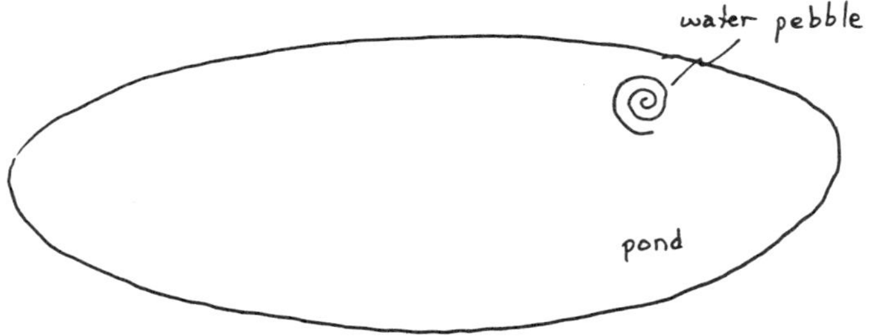

If the pond is initially still, the pebble creates spin within the pond as it relinquishes spin itself:

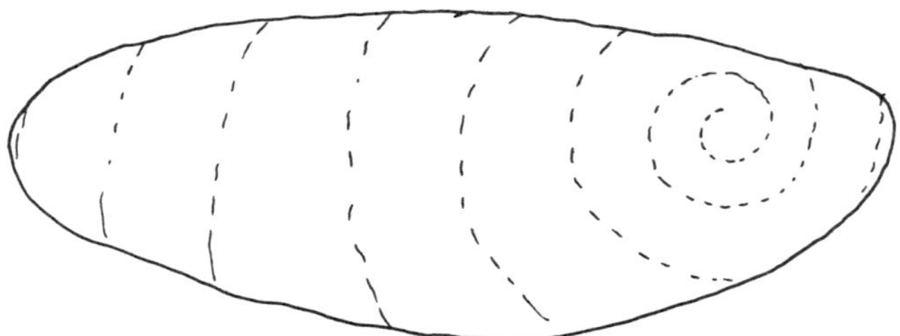

Now, if we have a group of water pebbles clustered together rather loosely to create the pond,

with all more or less maintaining *in situ* spin with minimal translation, the intrusion of one foreign pebble must affect them all. The instant the intruder hits the pond, it gives up linear motion and picks up spin. However the component pebbles, the pond's "nuclei", are essentially trapped; there's no way they can pick up linear motion or lose spin and still remain stable. The only way they can accommodate the excess is to destroy the pond, diffusing some of their spin elsewhere and accepting this intruding V_L. If they remain in their original configuration, the new V_L must find something else to disrupt.

Meanwhile our original pebble is assuming a much different form. If we track it from the moment it strikes the water to the point when one of the pebbles comprising the pond attempts to stabilize by striking something else, we would erroneously believe our first pebble was *directly* involved in all that. While the central pebbles scramble almost linearly to escape the highly unstable environment, our pebble leisurely continues giving up linear motion and picking up spin quite stably—so stably in fact its presence goes unnoticed in all the confusion. Thus while its linear energy is being converted to spin, the chaos at the other end is taken to mean energy is being produced.

If we think of the neutron striking the nucleus as a tiny whirlpool of water coming in contact with a larger, but more slowly moving whirlpool made up of other smaller ones, we can more easily envision the entire nuclear reaction as being part of the same thing. All the components in our system, regardless what happens to them, are merely different forms. The neutron can't "hit" or change anything because it's the same as everything else; all it can change is itself. However, by virtue of changing itself, the neutron indirectly changes the entire system. It functions not as an assailant so much as an initiator. The instant it enters the nuclear environment, its work is essentially done.

In a nuclear reaction the neutron functions like a backdoor intruder into an incredibly bad cocktail party where everyone is looking for an excuse to leave. At the front end of the room are numerous delicately balanced swinging doors that open at the merest touch; there's not much in the way of any physical barriers holding these less than compatible people together. At the opposite end of the room is a single door and it's through this our intruder suddenly charges—a crazy man, screaming, flailing his arms. Almost immediately a migration begins towards the other doors.

Obviously those out the doors first are the *least* likely to have any contact with the crazy man. Of all those in the room, they experience the minimum amount of interaction per unit time. If our crazy man is equivalent to a neutron in a fission reaction, what does this tell us about the neutron? If the opening of the back door and then the many front doors is likened to the first neutron's impaction with the nucleus and that of successively expelled particles with something else, we can see it represents a minimum amount of what the neutron actually does. If the first two party-goers out of the room run to the newspapers with their story, how accurately does it represent what actually occurs between each individual and our crazy man? If the reporter writes, "A crazy man cleared a party in less than a minute last night," how accurately does that describe the time frame of our demented man's effect on the group?

Suppose instead of being insane, our crazy turns out to be merely a rapidly twirling dancer. Once he enters the room, encounters a few guests and slows down, it becomes obvious he's neither crazy nor a threat. Those who are interested in dancing are drawn toward him. Those who aren't drift out the doors more or less leisurely, feeling this is sufficient reason to leave. Still others take advantage of the new arrival's behavior and begin dancing themselves.

In such a way our originally unstable group becomes quite stable with a core conversing with the dancer and the rest rhythmically flowing around them. The first two to leave now become the crazies who burst into the reporter's office with a tale that seems most alien to what those who chose to stay actually experienced. If our two evacuees happen to burst into another unstable boring party across the hall in their haste to leave the first, the process repeats itself. Only now, because twice as many crazies have come in from the back, four choose to leave from the front and we establish a chain reaction.

Let's take a closer look at our original crazy man. In the first place, we discover the time interval between his entry and the departure of the first two guests represents only a tiny fraction of that time he spends with the group. Furthermore if we evaluate the changes he and various members of the group experience during the total time, we find they're infinite. Obviously he was the reason the first two bored party goers burst out the doors; but he also conversed, danced, laughed, ate and drank with all combinations of others for hours. If we view each of these interactions as the creation of different forms of energy, how representative is his entrance and the exit of the first two others of the total?

Let's add one more wrinkle: Suppose our two cocktail party refugees get caught in traffic and it takes them three hours to file their report. The reporter's rendition and their account of the events will still be the same; the crazy man entered and the room emptied. Those who hear or read the account will envision the effect of the crazy person as most explosive and influential indeed. However we know that in that same period the effect of the crazy man was much less obstreperous but much longer lasting and powerful.

Such is the nature of our poor, maligned neutrons' place at the nuclear cocktail party. The total energy created isn't the total number of crazies bursting through one door and those out another. It's a measure of *all* the activities within the room, the majority of which might not be the least bit crazy. Having introduced a madman who turns out to be not mad at all, we see how the time designated between two events in a linear reaction doesn't necessarily mean that's all that occurs.

Let's review: The criteria established at our party are

- Bored guests looking for a reason to leave.
- Easily opened and closed doors.
- Sudden entry of a whirling individual from the opposite side of the room.

When we put these altogether, the following events occur:

- The dancer suddenly whirls into the room.
- Two bored party goers flee through opposite doors.
- They, in turn, wreak a little havoc with their entry and tale wherever they go.

From this we can readily see that it's unnecessary for the dancer to enter the room with enough force to knock the other two out; in fact, they have no physical contact at all. The other two are so ready to leave, his merely walking into the room is enough to precipitate their escape; all they need is someone to divert the others' attention while they make their exit.

Let's look at this in diagram form. Traditional linear thinking says neutron bombardment and subsequent energy release looks like this:

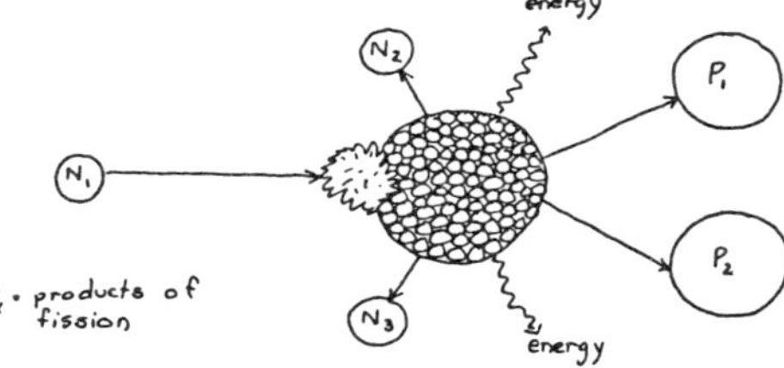

Rotational physics says the reaction is more like this:

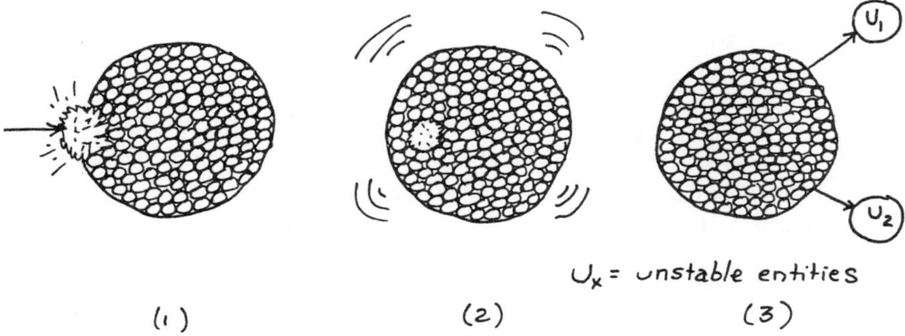

Linear physics says the *force* of the impact is what causes most additional neutrons to escape, disrupting the entire nucleus in the process.* Rotational physics says it's the addition of the neutron which creates or donates enough energy for the nucleus to be stable without the presence of unwilling, unstable "particles", enabling them to leave. Think of being engrossed in your favorite activity and suddenly being asked by your neighbor to babysit her unruly brood. You're part of their group and yet you aren't. The moment their mother returns, you're more than willing to leave; it's quite unnecessary for them to force you out.

At this point it would seem a conflict exists in current thinking regarding the use of radioactive materials. The mere instability of the starting mass should eliminate the need for any great force to release additional neutrons; and that is the case. From our discussion we can see how unstable or radioactive substances create a flow of loosely-bonded, escaped "particles" from one nucleus to another. One group of crazies migrates to an equally unstable party causing more potential crazies to leave that one and find another.

This in and of itself wouldn't be problematic if there were

---

*We are speaking here of fission caused by so-called "fast" neutrons. There are cases of both spontaneous fission and fission by thermal or "slow" neutrons, but they are not as prevalent.

sufficient room between nuclei. However, we know

- Our escapees want to get away from the group.
- The rest of the group is immediately drawn toward the new entrant.

Furthermore, these two conditions are simultaneous processes; the loss of energy out one door is instantaneously felt by the group and it gravitates toward the other door to fill that need. Therefore, the *only* way *anything* can escape via bombardment is if the energy in is *equal or greater* than that escaping. And that's inefficient to say the least!

In this principle we learned how to take a linear event or progression and break it down into an infinite number of points so we can assess its total potential. If we choose to see linearly, then that's all we see; if we choose to see multidimensionally, we increase our awareness a great deal. If all we see is a singular beginning, a singular process and a singular goal, that's *all* we can possibly experience. What we see is always what we get. However, what we see is a matter of choice; if we choose to see more or less, what is real to us is contracted or expanded by our choice.

This ends our discussion of the Twelfth Principle and our Twelve Principles of Energy. Many new concepts have been introduced and in so doing, it may initially appear that we've been severely critical of nuclear fission to a greater degree and fusion to a lesser one. It's not our intent to find fault or lay blame but rather to present logical arguments and examples showing why these processes should and could be brought to a quiet and peaceful close and replaced with something benign and beneficial.

17 | What Does This Mean to Me?

In this book we've seen how the simple rotational theory advanced in *Primer* applies to our beliefs about the creation and utilization of energy. To summarize the twelve principles, let's take each one and apply it to ourselves, knowing that's always the test for any theory's validity: If it doesn't hold true for me, it can't be true for anything else.

1. If energy is created at the smallest level, there's no waste. If we recognize that whatever we experience is what we want, whether we can immediately understand our reasons or not, we save ourselves a lot of unnecessary wear and tear. We don't do things for Mom, Dad, God and/or country: We do them only for ourselves and only we can change them if we don't like the results. Hooking others into our choices merely complicates things unnecessarily.

2. Process is less important than the result. There is, the old trappers say, more than one way to skin a cat. For the time-oriented supervisor to evaluate employee performance based on promptness is ludicrous; to judge parents by a set of fixed criteria rather than their relationship with their children has no meaning. To deny ourselves success because it's too easy, or the right to change because it denotes our failure in one of an infinite number of processes to reach the same goal, is counter-productive.

3. Heat-based reactions are wasteful and inefficient. How many times does our anger, ranting and raving ever accomplish anything beyond dissipating pent-up energy?

Rarely, if ever: Most of the time all we create is a lot of sound and fury. Then we must summon more energy to accomplish any change. In short, for all we put in, we get very little out.

4. Any chain reaction is unnatural. Any time we take an experience, lump it with similar experiences and increase or decrease our response to it exponentially, we create problems. An exponential increase in our response to others or events is perceived as overreaction; the exponential decrease leads to others feeling taken for granted or ignored. In both cases the individual event gives way to the process; it loses its identity and with it the opportunity to change.

5. Energy in one form can supply all the earth's needs. Although there are an infinite number of ways to accomplish any task, it is possible to use the same way again and again. As part of veterinary training, students are presented with a philosophical dilemma: Which is better—the practitioner who knows the workings of a few drugs inside out and backwards or the one with an armament of a thousand who knows a little about each one? Which approach we use isn't nearly so important as realizing (a) it's not the only approach, (b) others may use different approaches that work equally well for them, and (c) the use of any approach or form in no way negates our ability to use another.

6. The more specific the energy creation device, the more energy it wastes. The more definitions or specifications we build into a system, the more limiting it becomes. If we or others must act in a specific way to prove love, friendship, loyalty or intelligence, we cut ourselves off from all but an extremely narrow range of experience in those areas. Because we are either incapable of recognizing these other forms or choose to ignore them, they're essentially lost or wasted. Whenever we say "Only A can create B" we create a system whereby we can only have B if we have A. Such contingent if/then thinking ("If I have lots of money, then

I'll be happy." "If Frank loves me, he'll propose." "If Harry's a good employee, he'll be here on time.") makes it impossible for us to experience anything beyond that rigidly defined reality.

7. Whenever energy is stored, more energy is lost than gained. Philosophy, literature, music, mythology and our day-to-day activities are filled with examples of this most meaningful principle. Whenever we try to hang on to something because we value it, we invariably lose it. The rich man whose money merely collects more interest in various accounts gains nothing from his wealth. The person who never shows love for fear of being hurt winds up hurting anyhow. The person with knowledge who withholds it either out of lack of self-confidence or jealousy creates negative emotions for the self and nothing for anyone else. If we can store something, it means it has *little* value—we don't need it, we don't want to use it. If we don't need it or want to use it now, why waste all that time and energy accumulating an excess and storing it? It just doesn't make sense.

8. Mass is both created and destroyed. We each create and destroy our personal realities at will. If we feel strongly about someone or something, we may overlook what others consider quite obvious qualities while concentrating on those our friends consider minute or even imperceptable. It doesn't matter as long as we realize our "real" is neither absolute nor even necessarily someone else's real, and the same holds true for theirs. It's our willingness to accept the infinite variety of each individual's perceptions that creates the most stable world.

9. For every reaction in nature, some change in either mass or energy occurs. Nothing is insignificant; nothing is meaningless. Whatever we do has meaning, whether we choose to acknowledge it or not. This must be so if all is connected and unified.

10. Energy is neither created or destroyed. Love never dies, nor does our sense of who and what we are. Because these are energy forms, we can change our appearance, age, become "handicapped" in one way or another and it makes no difference—we always recognize the self and its infinite nature. Furthermore, we can accept change in others knowing their idea of mass and reality may differ from ours, but the basic intangible energies such as love and inspiration will always be present.

11. Any energy created above the level of the $_7\alpha$ isn't energy; it's an energy/mass combination. If we believe we do anything for any reason other than self-love, we delude ourselves and wind up denying our full potential. Isn't self-love selfish and ego-centric? Not in the least: If we each acknowledge we do everything, make any changes, because it's what we want, we give everyone and everything else the freedom to do the same. If we attribute any changes to energy or motivation provided by others, we start with an impure form. If our choice proves "good" we feel dependent on these others; if the choice turns out to be a "bad" one, we feel victimized by those same intermediate energy sources.

12. The total energy of any state is the sum of the energy of all its components at any given instant. If we use the purest form and don't store any, what is our personal energy total? The total is everything we've got. In other words, the system is designed to function at 100% all the time. No, this doesn't mean we run around frantically, constantly pushing ourselves to do more. It does mean we're totally committed to whatever it is we happen to be doing— whether it's reading a book, writing a report, watching the kids play Scrabble®, or enjoying a sunset. The full benefit, the total beauty of anything or anybody in this reality can only be appreciated if we decrease our resistance and let all our energy/love flow out and are willing to accept all

that flows in. If we have any reservations, we block that energy and deny ourselves its effects.

In the preceeding pages both nuclear fission and fusion came in for their share of criticism because of the anti-rotational, unnatural process underlying their creation. What does fission or fusion have to do with us as individuals? If all is connected, the mere existence of such processes, either in reality (fission) or theory (fusion) means they reflect prevalent human beliefs and/or conditions. Consider individuals who believe themselves at the mercy of other peoples, events, gods, whims, or energy. They're in a constant state of agitation, anticipating the unknown, not unlike radioactive materials. If the tension becomes great enough, one relatively small nudge is sufficient to create an explosive chain reaction. "All right, you creep, now you're going to pay for all the grief you've given me." Pow! In corporate board rooms where such extroverted chain reactions aren't acceptable, the body creates an internal one—hypertension, heart disease, stroke.

The fusion personality is a bit harder to fathom because by its very nature it's quite introverted. Any change is manifested inside—pressure increases as the system shrinks into itself. Think of those who seem to be so cool, so much in control nothing seems to bother them. While others are scrambling and emoting, these people are the proverbial rocks. However if that rock is really just a thick shell holding unstable material inside, the system is self-destructive. Because the unnatural energy is rigidly contained within the highly composed, resistant exterior, it has no external outlet; any changes manifested must occur within the body. Surely we needn't stretch our awareness too far to realize such is the nature of tissue-destructive processes such as cancer and degenerative diseases, including the auto-immune diseases like AIDS where the body literally attacks itself.

Are you a fission or fusion personality? Do you store things up for fear of hurting someone or being hurt yourself? Do you get so bogged down in process or hamper any creation with so many definitions or specifics, it turns out being less than what you wanted?

Do you believe you always get out of any relationship exactly what you put in, that there is no waste? Do you believe that whatever you give to or receive from any other is the purest form, freely given by choice?

Time and time again we've referred to the streaming effect and its philosophical corollary, "Insomuch as you do it unto the least, you do it unto me/all." What each one of us believes about our love of self and all else influences what we believe about the manner in which our homes are lighted and heated. If we don't believe we're worthy of love, if we know we don't love ourselves and can't count on love from others, control and storage will be very important to us. In an effort to control, we'll create specific definitions and processes which actually limit our effectiveness. If we don't want to accept responsibility for the reality each of us creates and destroys by choice, we attempt to create systems whereby we can *force* others to do what we want. In so doing we hope to convince ourselves the same is being done to us; the more control we feel we must exert over others, the *less* control we want to believe we have over ourselves.

Self-love, pure non-waste-producing self-energy, is the first step in solving our energy problems. Unless each one of us recognizes our individual ability to create and destroy mass at will as a function of how we choose to use our personal energy to create or destroy our perceptions, we can't make the necessary philosophical changes that would permit us to divest ourselves of current energy systems. It's no use bemoaning the evils of nuclear reactors and fossil fuels if we're afraid of the dark. If we feel safe and secure because we have savings accounts, pension plans and a year's supply of food in the pantry, chances are we also want on-demand, dependable energy sources with back-up systems to maintain them. Even though we know such systems are wasteful and depleting our natural resources, is our fear of the unknown sufficient to block out these negative effects? Unfortunately most of us must answer "Yes; I don't want to destroy my world but I want to be sure that energy will be there when I need it. I want physical proof that it's there waiting, and under my control."

Until each one of us has the courage to grant ourselves the full potential of our being, our energy problems can't be solved. To begin with, knowing the self is more than understanding the very least, the smallest familiar element in our worlds. To know the self—how we create mass from energy, the effects of process, storage and purity—is to be aware of all that is. This is the first step to making any lasting change and that change must start with what you know best:

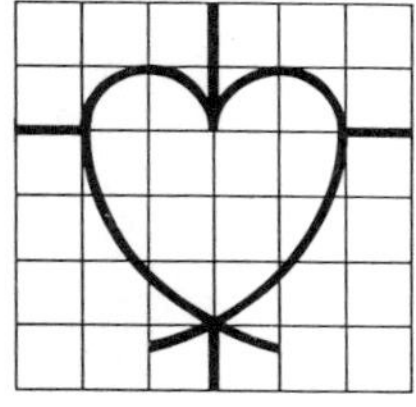

YOU.